The Original Collection of Math Contest Problems

Written By

Eric Xu

Michelle Chen

Victor Chen

Benjamin Chen

Jason Ye

Illustrated By

Lia Tian

Acknowledgements

We acknowledge all Orange County Math Circle volunteers who helped us in writing this book.

We thank all Orange County Math Circle donors for providing financial support for the publication of this book.

We express our sincere appreciation to all of our advisors, especially to Mrs. Wendy Li and Dr. James Li, for their great contributions to the Orange County Math Circle and their invaluable guidance during the development of this book.

We are also grateful for our parents' continuing support and encouragement.

Finally, our special thanks are extended to Professor Po-Shen Loh for writing the foreword, for his advice towards the book, and for his dedication to the Orange County Math Circle.

Contents

Foreword

Welcome, friend, to a book of problems! Does the world need more problems though? It appears to have plenty already.

Never fear—this book also has solutions! Indeed, the mission of this book is to help readers develop problem-solving superpowers that ultimately apply throughout all walks of life. Many of history's greatest breakthroughs arose from ingenious insights that cracked apparently hopeless challenges. This book provides a space to discover that joy of persevering from initial frustration to the *Aha!* moment.

Why math problems though? Although the ultimate goal may be to reverse climate change or end world hunger, mathematics provides the common language of logical analysis. Indeed, math is the art of thinking! The most challenging problems in this book derive their difficulty not from obscure facts and definitions, but from the creativity required to weave many elementary observations into sophisticated arguments. These fundamental and widely applicable problem solving skills will stay relevant long after the sines and cosines of high school Trigonometry have come and gone.

Where, then, do competitions enter the equation? The global math contest movement has built a vibrant culture around problems that truly bridge between the formulaic challenges of routine "schoolbook" exercises and real-world creative problem solving. The focus of contests—and indeed, of this book—is not to breed competition and division, but rather to unite a worldwide community of enthusiasts.

On that note, it is a singular pleasure to offer the Foreword for this book, as the Orange County Math Circle uniquely embodies that spirit of collaboration: it was founded by students, for students, as a service organization. Indeed, when I first heard about them, I simply *had* to visit in person. I discovered a group of students overflowing with positive energy, working to inspire others and share their talent with the world.

This book is a testament to their dedication. May it help you to develop your own superpowers as well. I then urge you to share them like the OCMC, to make tomorrow better than today!

Po-Shen Loh

Math professor, Carnegie Mellon University
National lead coach, USA International Math Olympiad Team
Co-Founder of expii.com, the Massive Open Online Collaboration

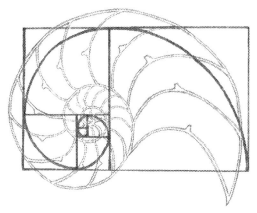

The pythagorean spiral is modeled off the golden ratio. This ratio, also known as the divine proportion, golden mean, or golden section, has connections to mathematical concepts such as continued fractions. In nature, this spiral is found in places such the spiral of a seashell.

Chapter 1

Algebra

Warm-up Problems

1. When a turtle wants to cross a river, he has to pay $4. If he crosses the river 5 times on Monday, how much did he pay on Monday? $20

2. Serge has a crush on Pam and decides to buy her 5 yards of yarn (and some roses). However, the yarn can only be bought in inches. How many inches of yarn does he need to buy?

180 in

3. If a bottle of sunscreen costs $8 and you have a coupon that gives a 25% discount, how many dollars do you pay if you use the coupon?

6 dollars

4. Allison took a series of tests during her freshman year. Her test scores were 92, 85, 87 and 100. However, instead of 100, her test score should have been a 98.5. What is the new median? Express your answer as a decimal rounded to the nearest tenth.

89.5

5. How many minutes are there in 20% of one day?

288 minutes

6. A fly eats 6 spiders every century. On average, how many spiders does the fly eat every year? Express your answer as a decimal rounded to the nearest hundredth.

7. Emily built a toothpick bridge that was strong enough to hold 15 pounds. How many $2\frac{1}{2}$ pound rats can sit on the bridge without it breaking?

8. If a dog house can hold up to 5 dogs, how many dogs can 25 dog houses hold?

125 dogs

9. If a frond grows at a rate of 2.5 centimeters per hour, how tall, in centimeters, is the fern after 2 days?

120 cm

2

10. If a bag of rice at 99 Ranch Market is $200, then how many dollars do 20 bags cost?

    $4000

11. You have 100 coins each worth the same amount of money. The total amount of money is $100. How much is each coin worth?

    10¢

12. Let us define a function $F(x)$ such that $F(x) = 0$ only when $x = 0$ or 1. What is $F(F(1))$?

Introductory Examples

1. For every bounce, a super-bouncy ball bounces to $\frac{5}{4}$ of the height of the previous bounce (or initial height for the first bounce). If it is dropped from a height of 16 feet, to what height in feet does the ball rise after the second bounce?

   Solution: After the first bounce, the ball will bounce to a height of $\frac{5}{4} \times 16 = 20$. After another bounce, the ball will rise to a height of $\frac{5}{4} \times 20 = \boxed{25}$ feet.

2. Compute the sum of the first 10 odd positive integers minus the sum of the first 10 even positive integers.

   Solution: The problem asks us to compute

   $$(1 + 3 + 5 + \cdots + 19) - (2 + 4 + 6 + \cdots + 20)$$

   We could find the two sums directly and then subtract. However, a more efficient method would be to group the expression and then evaluate like this:

   $$(1 - 2) + (3 - 4) + (5 - 6) + \cdots + (19 - 20)$$

   $$= (-1) + (-1) + (-1) + \cdots + (-1) = \boxed{-10}$$

3. What is the value of $(88)^2 - (12)^2$?

   Solution: $(88)^2 - (12)^2 = (88 - 12)(88 + 12) = (76)(100) = \boxed{7600}$.

4. Compute the value of $20^2 - 18^2 + 16^2 - 14^2 + \cdots + 4^2 - 2^2$.

   Solution: Rather than computing the value of the first ten even squares, we note that we have a lot of differences and

4

square numbers. This leads us to the idea of grouping every two consecutive terms in our expression:

$$(20^2 - 18^2) + (16^2 - 14^2) + \cdots + (4^2 - 2^2)$$

We can factor each grouping and simplify:

$$(20 + 18)(20 - 18) + (16 + 14)(16 - 14) + \cdots + (4 + 2)(4 - 2)$$

$$= 2(20 + 18 + 16 + 14 + \cdots + 4 + 2)$$

$$= 2 \cdot 2(10 + 9 + 8 + \cdots + 2 + 1)$$

$$= 2 \cdot 2 \cdot \frac{11 \cdot 10}{2} = \boxed{220}$$

5. In the magic square, the product of the numbers in each row and each column is the same. If all of the squares must be filled with positive integers, what is the value of a?

1		a
	3	
	4	

Solution: Let the middle square in the top row be labeled b. Then the product of the numbers in top row is $1 \cdot b \cdot a = ab$. The product of the middle row is $b \times 3 \times 4 = 12b$. Thus, $ab = 12b$. Since all the numbers have to be positive, we can divide through by b, giving us $a = \boxed{12}$.

1	b	a
	3	
	4	

6. One-third of Bob's current age is twice that of Bob's age 5 years ago. What is Bob's current age?

Solution: If x is Bob's current age, $x - 5$ must be his age 5 years ago, and $2(x - 5)$ would be twice his age 5 years ago. Since one-third of Bob's current age is $\frac{x}{3}$, we have

$$\frac{x}{3} = 2(x - 5) = 2x - 10 \implies x = 6x - 30$$

Solving for x, we get $x = \boxed{6}$.

7. Compute the value of x such that $\sqrt{8 + \sqrt{x + 30}} = \sqrt{3} + \sqrt{5}$.

Solution: To get rid of square roots, we square the equation:

$$8 + \sqrt{x + 30} = 3 + 5 + 2\sqrt{15} \implies \sqrt{x + 30} = 2\sqrt{15}$$

We now can square the equation again to get

$$x + 30 = 4 \cdot 15 = 60$$

so $x = \boxed{30}$.

8. Bernie the Bird is in the middle of the freeway. Car A is 50 miles due east of Bernie and is traveling at a rate of 150 miles per hour. Car B is traveling at a rate of 120 miles per hour and is 60 miles due west of Bernie. Car A will pass by Bernie and give food to him, but Car B will run over Bernie. After Bernie takes the food from Car A, how many minutes does Bernie have to get out of the street before Car B arrives?

Solution: Since Car A is 50 miles away traveling at 150 miles per hour, it will arrive at Bernie in $\frac{50}{150} \times 60 = 20$ minutes. Similarly, Car B will arrive at Bernie in $\frac{60}{120} \times 60 = 30$ minutes. As a result, Bernie has $30 - 20 = \boxed{10}$ minutes to get out of the street after receiving food from Car A.

9. Let $a_1, a_2, a_3, \ldots$ be a sequence of numbers such that $a_1 = 1$ and $a_n = n \cdot a_{n-1} + 1$ for $n \geq 2$. What is the value of a_4?

Solution: Since we only want the fourth term, we can simply brute force this:

$$a_2 = 2 \cdot a_1 + 1 = 3$$
$$a_3 = 3 \cdot a_2 + 1 = 10$$
$$a_4 = 4 \cdot a_3 + 1 = \boxed{41}$$

10. Harry's magic wand grows at a rate of 3 centimeters per minute. If the wand is initially 30 centimeters long, how many centimeters long will it be one hour later?

    Solution: There are 60 minutes in one hour, so in hour, the wand grows $3 \times 60 = 180$ centimeters. Initially, it was 30 centimeters long, so after one hour, it is $30 + 180 = \boxed{210}$ centimeters long.

11. Solve the following equation for $\sqrt{x}$:

$$(2014 + 2014) + (2014 - 2014) + x = 2014 \cdot 2014 + \frac{2014}{2014}$$

    Solution: We see that 2014 appears a lot in the equation, so we can let $y = 2014$. We then have

$$(y + y) + (y - y) + x = y^2 + \frac{y}{y}$$

$$2y + x = y^2 + 1$$
$$x = y^2 - 2y + 1 = (y - 1)^2$$

    As a result, the square root of x is $y - 1 = 2014 - 1 = \boxed{2013}$.

12. A CD spins at a rate of 27 revolutions per a minute. How many minutes will it take for the CD to spin 621 revolutions?

    Solution: We convert from revolutions to minutes using our given rate:

$$621 \text{ revolutions } \times \frac{1 \text{ minute}}{27 \text{ revolutions}} = \boxed{23} \text{ minutes.}$$

13. Bryan has a bag filled with candies. Unfortunately, every night, Ryan steals 70% of the number of candies left in the bag. If after three nights, Ryan has taken 1946 candies, how many candies does Bryan have left?

Solution: Let n be the number of candies Bryan had originally. After one night, Bryan will have $0.3n$ candies left because Ryan took $0.7n$ of them. Likewise, after two nights, Bryan has 0.3^2n candies left, and after three nights, he will have 0.3^3n candies left. Since Ryan took 1946 candies altogether, $0.3^3n = n - 1946$. We have $0.027n = n - 1946$, so $0.973n = 1946$. Dividing the equation by 0.973 gives us $n = 2000$. It follows that Bryan has $2000 - 1946 = \boxed{54}$ candies left.

14. There are 16 coins in a bank. If the coins are all nickels and dimes, and their total value is $1.05, how many nickels are there?

Solution: Let n be the number of nickels and d be the number of dimes. There are 16 coins in the bank, so $n + d = 16$. The total number of cents is 105, implying that $5n + 10d = 105 \implies n + 2d = 21$. Subtracting $n + d = 16$ from this one produces $d = 5$. It follows that $n = 16 - d = 16 - 5 = \boxed{11}$.

15. The number 66 is written as the sum of 2 numbers. One of these numbers is 3 more than twice the other number. Find the larger of the two numbers.

Solution: If we let a be the larger number and b be the smaller number, the sentence "One of these numbers is 3 more than twice the other number" can be expressed as $a = 3 + 2b$. Furthermore, the sum of the two is 66, so $a + b = 66$. Substituting $a = 3 + 2b$ into $a + b = 66$, we get $3 + 2b + b = 66$. Solving for b, we have $b = 21$, making $a = 66 - 21 = \boxed{45}$.

16. Denote $\sqrt[n]{a}$ as the nth root of a. What is the value of n if $\sqrt[n]{112} = 2\sqrt[n]{7}$?

Solution: Rewriting the roots as exponents, we get

$$112^{\frac{1}{n}} = 2 \cdot 7^{\frac{1}{n}}$$

Dividing both sides by $7^{\frac{1}{n}}$, we have

$$\frac{112^{\frac{1}{n}}}{7^{\frac{1}{n}}} = 2 \implies \left(\frac{112}{7}\right)^{\frac{1}{n}} = 2$$

or $16^{\frac{1}{n}} = 2$. Since $2^4 = 16$, $16^{\frac{1}{4}} = 2$, so $n = \boxed{4}$.

17. The price of a jacket was 80% of its original price during a sale. Michelle bought the jacket using a coupon for 15% off the sale price. If she paid with a 50 dollar bill and got \$16 as change, how many dollars did the jacket cost originally? Assume no tax.

    Solution: Let x be the original price of the jacket. We know that before the coupon, the jacket costs $0.8x$ dollars. If the coupon gives 15% off the sale price, then after the coupon is applied, the jacket now costs $0.85 \cdot 0.8x$ dollars. Since Michelle paid with a 50 dollar bill and got 16 dollars in change, the jacket must be worth $50 - 16 = 34$ dollars after the coupon. As a result, we have $0.85 \cdot 0.8x = 66$, so $x = \dfrac{66}{0.85 \cdot 0.8} = \boxed{50}$.

18. The length of a rectangular picture is 3 times its width. The picture is surrounded by a frame which is 4 inches wide. If the perimeter of the outside of the frame is 96 inches, what is the length of the picture in inches?

    Solution: Let w be the width of the picture. The first sentence of the problem tells us that the length of the picture is $3w$. Furthermore, if the frame is four inches wide, the dimensions of the frame must be $w + 8$ and $3w + 8$. Since the perimeter of the frame is 96 inches, we have

    $$2(w + 8 + 3w + 8) = 96$$

    Dividing both sides by 2 and simplifying, we get

    $$4w + 16 = 48 \implies 4w = 32 \implies w = 8 \implies l = 3(8) = \boxed{24}$$

19. The price of a field trip was evenly split among 17 students and 3 teachers. Then 4 of the students said that they couldn't go, a new teacher joined, and 7 new students joined. The price was again evenly split, and the average price dropped 60 cents. How many dollars did the field trip cost?

Solution: Notice that there is no distinction between the price of a student and the price of a teacher. Thus, we only need to keep track of the total number of people. Initially, we have $17 + 3 = 20$ people. At the end, we have $20 - 4 + 1 + 7 = 24$ people. Therefore, if we let p be the average price initially, we have

$$20p = \text{ cost of the trip } = 24(p - 0.6)$$

Rearranging terms, we have $14.4 = 4p$, so $p = 3.6$. It follows that the total trip costs $20(3.6) = \boxed{72}$ dollars.

20. At $8 : 30$ AM, Linda starts running around a circular 200 meter path at 8 meters per second. Eight minutes later, Larry begins running at the same starting place as Linda and in the same direction at 10 meters per second. How many meters will Larry run to first catch up to Linda?

Solution: Since we know that Larry begins running eight minutes after Linda starts, we can first calculate where Linda is when Larry begins running. The path has a circumference of 200 meters and Linda runs at 8 meters per second, so she takes $\frac{200}{8} = 25$ seconds to run a lap.

There are $8 \times 6 = 480$ seconds before Larry starts to run. Every 25 seconds, Linda arrives at her starting place again. Therefore, she will be $480 \equiv 5 \pmod{25}$ seconds into her current lap when Larry begins. Running at 8 meters per second, Linda will be $5 \times 8 = 40$ meters ahead by the time Larry begins. Since Larry runs at 10 meters per second, each second Larry decreases the gap between Linda and himself by 2 meters. The gap is initially 40 meters, so it takes 20 seconds for Larry to catch up, meaning he has ran $20 \times 10 = \boxed{200}$ meters.

21. How many non-congruent and non-equilateral triangles with a perimeter of 120 units have integer side lengths that form an arithmetic sequence?

Solution: We have a three-term arithmetic sequence and the terms sum to 120. When dealing with an arithmetic sequence of odd length, computation is easier when we let the "starting point" of the sequence be the middle term. In other words, let the middle side be a and let the common difference be d. It follows that the shortest side is $a - d$ and the longest is $a + d$. Then $a - d + a + a + d = 120$, so $a = 40$.

Since the side lengths of the triangle must be positive, $a - d > 0$, so $d < 40$. Furthermore, the lengths have to satisfy the Triangle Inequality, so $a - d + a > a + d \implies 40 > 2d$, or $d < 20$. We are also looking at non-equilateral triangles, so $d > 0$.

As a result, for every integer d between 1 and 19 inclusive, we obtain a unique triangle satisfying the conditions. Because of this, there are $\boxed{19}$ such triangles.

22. In the arithmetic sequence $3, 6, 9, x, 15, y, 21...$ what is the value of the expression $3x - 2y$?

Solution: From the first three terms, we see that the common difference is 3. Then $x = 9 + 3 = 12$ and $y = 15 + 3 = 18$. Hence, $3x - 2y = 3(12) - 2(18) = \boxed{0}$.

23. Solve for x in the following equation given that x is non-zero:

$$\frac{3x}{2x} + \frac{3}{4} = \frac{3x + 3}{2x + 4}$$

Solution: Because $x \neq 0$, we can cancel out the factors of x in $\frac{3x}{2x}$. Then we can simplify the left side by combining denominators to get

$$\frac{9}{4} = \frac{3x + 3}{2x + 4}$$

Cross multiplying, we get

$$9(2x + 4) = 4(3x + 3) \implies 18x + 36 = 12x + 12$$

Finally, we solve for x, giving us $6x = -24$, or $x = \boxed{-4}$.

24. Bradley's first four math test scores are 80, 77, 98, and 85. If math test scores are integer values from 0 to 100 inclusive, how many possible scores can Bradley get on his fifth test so that the median of all five of his tests is exactly 85?

Solution: The median of a set of values is the middle number when the values are sorted from least to greatest. Currently, Bradley's test scores can be ordered as $\{77, 80, 85, 98\}$. If 85 was the median of five test scores, 85 would have to be the third number in the sorted set. If we placed a score less than 85 into the set, 85 would become the fourth number and would not be the median. (For example, if the fifth score is 82, the set would become $\{77, 80, 82, 85, 98\}$.) Thus, Bradley's fifth score must be 85 or greater. The maximum score is 100, so there are $100 - 85 + 1 = \boxed{16}$ possible scores.

25. An arithmetic sequence satisfies the property that the first term of the sequence is three times the common difference. If the sum of the first ten terms is 5625, what is the common difference of the sequence?

Solution: To start off, we can write the given property as an equation. If a is the first term and d is the common difference, we have $a = 3d$. We also know that the sum of the first ten elements is 5625. The tenth term is $a + 9d$, so the sum of the first ten terms is $\dfrac{(a + a + 9d)10}{2} = 5(2a + 9d)$. Substituting in our relationship between a and d, we have

$$5(2(3d) + 9d) = 5625 \implies 75d = 5625$$

so $d = \boxed{75}$.

26. The sum of the first nine terms of an arithmetic sequence is 45. What is the fifth term of the arithmetic sequence?

Solution 1: Let the first term of the sequence be a and the common difference be d. Then we can express the sum of the first nine terms as

$$\frac{(a + a + 8d)9}{2} = \frac{(2a + 8d)9}{2} = 9(a + 4d) = 45$$

As a result, $a + 4d = 5$. We are asked to find the fifth term, which is exactly $a + 4d$! Thus, the fifth term is equal to $\boxed{5}$.

Solution 2: Alternatively, since we want to find the fifth term, we can set a to be the fifth term and d to be the common difference. Then our first nine terms are

$$a - 4d, a - 3d, a - 2d, a - d, a, a + d, a + 2d, a + 3d, a + 4d$$

Summing up these 9 terms, we see that the common difference, d, cancels out, leaving us with $9a = 45$, so $a = \boxed{5}$.

27. In Farmer Fred's barn, there are pigs and cows. If Farmer Fred sells 5 pigs, he will have twice as many cows as pigs. But if Farmer Fred instead buys 17 more cows, he will have 3 times as many cows as pigs. How many cows does Farmer Fred have currently?

Solution: To solve this problem, we translate the English sentences into mathematical equations. Let p be the number of pigs Farmer Fred has and c be the number of cows. The sentence "If Farmer Fred sells 5 pigs away, he will have twice as many cows as pigs" tells us that $2(p - 5) = c$. Likewise, the sentence "But if Farmer Fred instead buys 17 more cows, he will have 3 times as many cows as pigs" can be written as $c + 17 = 3p$.

We now have two equations and two variables, so we can solve the system. We can substitute $c = 2(p-5)$ into $c+17 = 3p$ to get $2(p - 5) + 17 = 3p$. Solving for p, we have $2p - 10 + 17 = 3p \implies p = 7$. As a result, Farmer Fred has $2(7 - 5) = \boxed{4}$ cows.

13

28. Barbara the Barber can cut a person's hair in twenty minutes for ten dollars and gel a person's hair in five minutes for two dollars. After working for one hour, Barbara has altogether cut or gelled six people's hair. If nobody's hair was both cut and gelled, how much money did Barbara make?

Solution: Let x be the number of people who had their hair cut by Barbara, and y be the number of people who had their hair gelled by Barbara. Since there were six customers total, $x + y = 6$. Then using the times given in the problem, we have $20x + 5y = 60$, which can be simplified to $4x + y = 12$.

Subtracting $x + y = 6$ from $4x + y = 12$, we get $3x = 6$, so $x = 2$. Plugging our value for x into $x + y = 6$ produces $y = 4$. Hence, Barbara will make $2(10) + 4(2) = \boxed{28}$ dollars.

29. Given two distinct real numbers x and y, define $x \& y$ as $\dfrac{2x^2 - 3xy + y^2}{x - y}$. What is the value of $1\&2 + 2\&3 + 3\&4 + \cdots + 9\&10$?

Solution: The operation $x \& y = \dfrac{2x^2 - 3xy + y^2}{x - y}$ seems pretty scary right now, so instead of plugging in numbers into it a bunch of times, we can try to simplify the big fraction. Note that $2x^2 - 3xy + y^2$ can be factored as $(2x - y)(x - y)$. Since we know that $x \neq y$,

$$x \& y = \frac{(2x - y)(x - y)}{x - y} = 2x - y$$

This is much easier to work with! Our expression then simplifies to

$$1\&2 + 2\&3 + 3\&4 + \cdots + 9\&10$$
$$= (2(1) - 2) + (2(2) - 3) + \cdots + (2(9) - 10)$$
$$= 2(1 + 2 + \cdots + 9) - 2 - 3 - \cdots - 10 = 2(45) - 54 = \boxed{36}$$

14

30. Frank and Fiona are running laps on a circular track. One lap around the track is 400 meters. Frank runs at 3 meters per second and Fiona runs at 2 meters per second. If Frank and Fiona start running from the same spot but run around the track in opposite directions, how many times will they pass each other after five minutes?

Solution: We can first calculate how much time passes between Frank and Fiona's first and second time meeting each other. Observe that when they meet again, the sum of the distances they traveled together is equal to a lap on the track. Thus, if t is the amount of time has passed since they last met, we have $3t + 2t = 400$, so $t = 80$ seconds. Over five minutes, or 300 seconds, Frank and Fiona will have met $\dfrac{300}{80} = 3\dfrac{3}{4} \implies \boxed{3}$ times (They have not yet crossed for the fourth time).

31. If $x + y = 8$ and $xy = 15$, compute $x^2 + y^2$.

Solution: Squaring $x + y = 8$ yields $x^2 + 2xy + y^2 = 64$ and since $xy = 15, 2xy = 30$ and $x^2 + y^2 = \boxed{34}$.

Difficult Problems

1. Compute $\frac{x^8+1}{x^4}$ if $\frac{x^2+2x-1}{x} - 2 = 5$.

   Solution: We can first try to solve for x because we are given a quadratic:

   $$\frac{x^2 + 2x - 1}{x} - 2 = 5 \implies x^2 - 5x - 1 = 0$$

   However, the roots of the quadratic are irrational, so we don't want to compute x^4 or x^8.

   Instead, we can try to regroup the numerator, so $\frac{x^8+1}{x^4}$ becomes $x^4 + \frac{1}{x^4}$ and $\frac{x^2+2x-1}{x} - 2$ becomes $x + 2 - \frac{1}{x} - 2 = x - \frac{1}{x}$. Now, we have $x - \frac{1}{x} = 5$. We need x to a fourth power, so we can first square the equation:

   $$\left(x - \frac{1}{x}\right)^2 = x^2 - 2 + \frac{1}{x^2} = 25 \implies x^2 + \frac{1}{x^2} = 27$$

   We can then square it again to get:

   $$x^4 + 2 + \frac{1}{x^4} = 27^2 = 729$$

   so the answer is $\boxed{727}$.

2. A rectangle has a perimeter of 48 centimeters and an area of 40 square centimeters. Find the sum of the squares of the length and width of the rectangle.

   Solution: Let l be the length and w be the width. Since the perimeter is 48, we have $2(l + w) = 48 \implies l + w = 24$. Moreover, the area is 40, so $lw = 40$. Observe that we want to find the value of $l^2 + w^2$. Rather than solving the two equations for l and w and then computing $l^2 + w^2$, we can square $l + w = 24$:

   $$(l + w)^2 = l^2 + w^2 + 2lw = 24^2 = 576$$

   Since $lw = 40$, we have $l^2 + w^2 + 2(40) = 576$, so $l^2 + w^2 = \boxed{496}$.

16

3. The sum of two numbers is 192 and the sum of their reciprocals is 8. What is the product of the two numbers?

Solution: Let a and b be the two numbers. We have $a + b = 192$ and $\frac{1}{a} + \frac{1}{b} = 8$. We want to find the value of ab. Instead of solving for a and b, we first combine the denominators in the second equation to get

$$\frac{a + b}{ab} = 8$$

We now can substitute $a + b = 192$ into this new equation to get

$$\frac{192}{ab} = 8 \implies ab = \frac{192}{8} = \boxed{24}$$

4. Albert, Bernard, Celine, and Eduardo discuss their heights. They discover Celine's height is 21 inches shorter than $\frac{3}{2}$ of Bernard's height. Albert is $\frac{4}{3}$ as tall as Celine, and Eduardo's height of 56 inches is the average of the two tallest heights (not including his). What is Bernard's height (in inches), given that everyone is taller than 43 inches?

Solution: Let a denote Albert's height, b denote Bernard's height, and so on. We know that Albert must be taller than Celine. If Bernard was taller than Celine, then

$$\frac{3}{2}b - 21 = c < b \implies \frac{1}{2}b < 21 \implies b < 42$$

which is a contradiction since we know everyone is taller than 43 inches. Thus, the tallest two people other than Eduardo are Celine and Albert. Thus,

$$56 = \frac{a + c}{2} = \frac{\frac{4}{3}c + c}{2} = \frac{7}{6}c$$

Multiplying both sides by sixth-sevenths, we get $c = 48$. We now can solve for b:

$$\frac{3}{2}b - 21 = 48 \implies \frac{3}{2}b = 69$$

$$b = \boxed{46}$$

5. If $\dfrac{x^4 + 1}{x^2} = 14$ and $x > 0$, what is the value of $\dfrac{x^6 + 1}{x^3}$?

Solution: If we break up the numerators, we have $x^2 + \frac{1}{x^2} = 14$, and we want to find $x^3 + \frac{1}{x^3}$. If we knew what $x + \frac{1}{x}$ was, we could cube that to get an expression containing $x^3 + \frac{1}{x^3}$. Unfortunately, we only have $x^2 + \frac{1}{x^2} = 14$. Still, we can manipulate that equation by adding 2 to both sides and completing the square:

$$x^2 + 2 + \frac{1}{x^2} = 16$$

$$\left(x + \frac{1}{x}\right)^2 = 16$$

Thus, $x + \frac{1}{x} = 4$ $(x > 0)$. Now, cubing the equation, we have

$$x^3 + \frac{1}{x^3} + 3\left(x + \frac{1}{x}\right) = 4^3 = 64$$

$$x^3 + \frac{1}{x^3} + 3(4) = 64$$

Hence, our answer is $64 - 12 = \boxed{52}$.

6. If $x = 2\sqrt{3\sqrt{2\sqrt{3\sqrt{2...}}}}$, what is the value of $x^{\frac{3}{2}}$?

Solution: Note that since the square roots repeat forever, we can write

$$x = 2\sqrt{3\sqrt{x}}$$

because the infinite square roots repeat. Squaring both sides and simplifying, we get

$$x^2 = 4 \cdot 3\sqrt{x} = 12x^{\frac{1}{2}}$$

$$x^{\frac{3}{2}} = \boxed{12}$$

18

7. Between 7:00 and 8:00, the minute hand and hour hand will form a 25 degree angle twice, at m and n minutes after 7:00. What is the value of $m + n$? Express your answer as a common fraction.

Solution: We first consider how many degrees each hand moves in one minute. Since the minute hand moves 360 degrees in 60 minutes, it moves at a rate of 6 degrees per minute. The twelve numbers (1 through 12) on the clock are evenly spaced, and the hour hand moves from one number to the next one in 60 minutes. The number of degrees between two adjacent numbers is $\dfrac{360}{12} = 30$. Thus, the hour hand moves at a rate of half a degree per minute.

At 7:00, the hour hand is $7(30) = 210$ degrees away clockwise from the '12'. Let x be the number of minutes past 7 such that the hands form a 25 degree angle. We then write

$$\left| 6x - \left(\frac{x}{2} + 210 \right) \right| = 25$$

$$\left| \frac{11x}{2} - 210 \right| = 25$$

To remove the absolute value, we take two cases. We either have

$$\frac{11x}{2} - 210 = 25 \implies \frac{11x}{2} = 235 \implies x = \frac{470}{11}$$

or

$$210 - \frac{11x}{2} = 25 \implies \frac{11x}{2} = 185 \implies x = \frac{370}{11}$$

Adding both values of x, we get that $m + n = \dfrac{470}{11} + \dfrac{370}{11} = \boxed{\dfrac{840}{11}}$.

8. Every day, Pat drives to work from her house. She first must drive 24 miles to the freeway, averaging a rate of 30 mph, and then drives 60 mph on the freeway to her workplace. Ruth takes the train directly to the same workplace, but the train

takes off one hour after Pat leaves her house. Both Pat and Ruth arrive at the workplace at the same time. If the train runs at a rate of 100 mph and Ruth lives 30 miles from the workplace, how far is Pat's house from the workplace?

Solution: We can relate Pat's route and Ruth's route using time because we know that both of them reach the workplace at the same time. We are given that Ruth travels 30 miles at 100 mph on a train, so she takes $\dfrac{30 \text{ mi}}{100 \frac{\text{mi}}{\text{hr}}} = \dfrac{3}{10}$ hours to get to work. On the other hand, Pat takes $\dfrac{24}{30} = \dfrac{4}{5}$ hours to get to the freeway, plus an additional $\dfrac{d}{60}$ hours on the freeway to reach the workplace, where d is the distance Pat drives on the freeway. So Pat takes $\dfrac{4}{5} + \dfrac{d}{60}$ hours to get to work. We are also given that Pat leaves one hour earlier than Ruth. Then Pat's total time minus Ruth's total time should be equal to 1. This gives us the equation

$$\frac{4}{5} + \frac{d}{60} - \frac{3}{10} = 1$$

We can multiply the equation by 60 on both sides, which leads to

$$48 + d - 18 = 60$$

Solving for d, we have $d = 30$ miles. Adding on the distance Pat drives from her house to the freeway, we get that her house is $30 + 24 = \boxed{54}$ miles from the workplace.

9. Ed arrives at a barber shop at 4:20 PM to get a haircut. When he leaves the shop, he notices on his analog watch that the minute and hour hand point to diametrically opposite points. If Ed was in the barber shop for less than one hour, how many minutes did Ed stay in the shop? Express your answer as a decimal rounded to the nearest tenth.

Solution: We first consider how many degrees each hand moves in one minute. Since the minute hand moves 360 degrees in 60 minutes, it moves at a rate of 6 degrees per minute.

The twelve numbers (1 through 12) on the clock are evenly spaced, and the hour hand moves from one number to the next one in 60 minutes. The number of degrees between two adjacent numbers is $\frac{360}{12} = 30$. Thus, the hour hand moves at a rate of half a degree per minute.

Since we know the hour hand and minute hand form a diameter on the clock, we can try to find the number of degrees each hand has moved since 4:00 PM. Let m be the number of minutes that has passed since 4:00 PM when Ed leaves the shop. The minute hand has traveled $6m$ degrees. On the other hand, the hour hand has traveled $\frac{m}{2}$ degrees from the '4' on the clock. Furthermore, the '4' is $4 \times 30 = 120$ degrees away from the '12', so the hour hand is $\frac{m}{2} + 120$ degrees away from the '12'.

Hence, we can write

$$6m - \left(\frac{m}{2} + 120\right) = 180$$

$$6m - \frac{m}{2} - 120 = 180$$

$$\frac{11m}{2} = 300$$

$$m = \frac{600}{11}$$

As a result, Ed has been in the shop for $\frac{600}{11} - 20 \approx \boxed{34.5}$ minutes.

10. When it is between 2:00 PM and 3:00 PM, Becky looks at an analog clock, and she discovers that the hour hand lies exactly on top of the minute hand. What time is it? Round to the nearest minute.

Solution: If the hour hand lies exactly on top of the minute hand, both hands must be the same number of degrees away from the '12' on the clock. Since the minute hand moves 360 degrees in 60 minutes, it moves at a rate of 6 degrees per minute. The twelve numbers (1 through 12) on the clock are evenly spaced, and the hour hand moves from one number to the next one in 60 minutes. The number of degrees between two adjacent numbers is $\dfrac{360}{12} = 30$. Thus, the hour hand moves at a rate of half a degree per minute.

If m is the number of minutes after two o'clock, then the minute hand has moved $6m$ degrees from the '12'. On the other hand (pun intended?), the hour hand has moved $60 + \dfrac{m}{2}$ degrees from the '12'. Since the two hands overlap, we have

$$6m = 60 + \frac{m}{2} \implies \frac{11m}{2} = 60$$

As a result, $m = \dfrac{120}{11} \approx 11$. The time is $\boxed{2:11 \text{ PM}}$.

11. Evaluate $\sqrt{90 + \sqrt{90 + \sqrt{90 + \cdots}}}$.

Solution: To deal with the infinite square roots, we can first set $x = \sqrt{90 + \sqrt{90 + \sqrt{90 + \cdots}}}$. We then notice that everything under the second square root is equal to x as well because the square roots continue forever! Thus, we have

$$x = \sqrt{90 + x} \implies x^2 = 90 + x \implies x^2 - x - 90 = 0$$

Factoring the quadratic, we have $(x-10)(x+9) = 0$, so $x = 10$ or $x = -9$. Obviously, $x = -9$ is an extraneous solution, so we conclude that our expression evaluates to $\boxed{10}$.

Additional Problems

1. If $3x + 4y = 13$ and $4x + 3y = 22$, compute $x + y$.

2. Given lines $3x + 4y = 12$, $6x + 8y = 17$ and $x + y = 13$, how many points (x, y) lie on exactly two of these lines?

3. Victor is sharing his baguettes. If he has a supply of nine hundred baguettes, and gives out 1 during the first minute, 3 during the second minute, 5 during the third minute, and so on, how many minutes will it take for him to run out of baguettes?

4. Eight students graduate from Orange County Math Circle. 17 more enter from overseas due to unsatisfactory teaching. Another 12 leave since they were assigned too much homework. There are now 26 members. How many people were in Orange County Math Circle originally?

5. In Calculus class, tests are worth 85% of the grade and the final is worth 15%. If William has an 86% in the test category, what grade does William need on his final to maintain a 70% in the class?

6. If you walk for 45 minutes at a rate of 4 miles per hour and then run for 30 minutes at a rate of 10 mph, how many miles will you have gone at the end of the one hour and 15 minutes?

7. A shirt is on sale for $22. The shirt was discounted by 12%. What was the original price of the shirt in dollars?

8. Define a function $a <> b = a^b + b^a$. What is $(3 <> 2)(1 <> -1)$?

9. Marx is walking down a hallway when he hears a fluffy unicorn behind him. Both Marx and the unicorn begin running at the same time. Marx runs at 3 meters per second while the unicorn gallops at 10 meters per second. If Marx was trampled by the unicorn after 10 seconds, how far, in meters, was the unicorn behind Marx when Marx heard her?

10. A sequence is defined as follows: $a_1 = 1$, $a_2 = 0$ and $a_k = a_{k-2} + a_{k-1}$ for $k \leq 3$. What is a_5?

11. If 5 times a number is 4, what is 200 times the reciprocal of the number?

12. Michelle has a pet dog weighing 20 pounds. It grows 0.7 pounds every week. It loses 0.1 pounds every day. How many pounds does it weigh after 7 weeks?

13. If the average ACT score at University High School is 36 out of a group of 50 students and the average ACT score at Corona Del Mar is 39 out of a group of 100 students, what is the average ACT score of both schools?

14. A CAM costs $2000 while a CP costs $30,000. How many CAMs can you buy for the price of a CP?

15. Jessica is taking a Spanish class at school. This semester, she has received test scores of 90, 84, 96, and 98. She wishes to receive an A this semester, which requires a test average of 93. Compute the minimum score that she must receive for her last test in order to receive an A.

16. Two people are running toward each other at 5 miles per hour in opposite directions on adjacent sidewalks. A bird decides to have some fun and flies between the two people at 18 miles per hour until the runners meet. It begins flying when the runners are 20 miles apart. What is the total distance the bird will fly?

17. Jon begins eating a box of 50 cookies at a rate of 2 cookies per minute. Wishing to join in the fun after watching Jon eat for 5 minutes, Allie shares Jon's cookies at a rate of 3 cookies per minute. Jon is feeling a little full, so when Allie begins eating cookies with him, he slows down to a rate of 1 cookie per minute. How many cookies has Allie eaten when they finish the box?

18. At 2:30 PM what is the smaller angle is formed by the hour and minute hand?

19. If $a = -1$, compute the largest number in the set

$$\{-3a, 4a, \frac{24}{a}, a^2, 1\}.$$

20. Dorky buys a shirt for $20 dollars and is charged an additional 6.5% sales tax. Lucky buys the same shirt but is charged an additional 6% sales tax. How many more cents does Dorky pay than Lucky?

21. Mr. Frahit receives a 10% raise every year. His salary after four such raises has gone up by what percent?

22. If the French Army retreats 3 times faster than than the British Army, and the British Army retreats 9 times faster than the German Army, how fast (in meters per second) do the French run if the Germans retreat at a rate of 8 meters per second from Pakistan?

23. This year, 700,000 people took the AMC. 5% of these passed into AIME. 10% of the people who take the AIME can take the USAMO. 0.2% of USAMO takers pass into IMO. How many people passed into the IMO this year?

24. Sale prices at the OCMC Outlet Store are 50% below original prices. On Saturdays an additional discount of 20% off the sale price is given. What is the Saturday price (in dollars) of a coat whose original price is $180?

25. Jane went to the zoo and saw that there were a lot of visitors that day. Since there was a long wait before she could enter the zoo, she counted tires, sedans, and motorcycles in the parking lot. If she counted a total of 100 vehicles and 340 tires, how many motorcycles did she see?

26. Mary is shopping at her favorite mall. In the first store she shops in, Mary spends 20% of her money. Then, she spends 25% of her remaining money on fancy pens and pencils. Before she goes home, Mary eats lunch, saving 80% of the money she had left. When she arrives at her house, she finds that she has $24 remaining in her purse. How much money did she bring to the mall?

27. Let $a \bowtie b = \dfrac{2a + b}{a^2 + b^2}$. Compute $3 \bowtie (2 \bowtie 1)$.

28. A poodle usually drinks water at a rate of 1 liter per 7 minutes. Since the poodle is very thirsty, its drinking rate triples. If we leave six-sevenths of a liter out for the poodle, in how many minutes will it finish?

29. I am a number between 1 and 100. You get the same number if you multiply me by 4 or add 198 to me. What number am I?

30. A dog runs toward a ball 30 feet away. When the dog is halfway there, how much farther does he have to run to get the ball and get back to his starting point (in feet)?

31. If the beaver slaps his tail on water 4 times over 5 seconds, then pauses for 5 seconds, then repeats the cycle, how many times does he slap his tail on the water in one minute?

32. The product of all 4 sides of a square is 1296. What is the sum of all four of the square's sides?

33. John has some cookies. He gives half of them to his siblings, one-third of them to his parents, and one-twelfth to his friend. He eats the only cookie left. How many cookies did he start with?

34. If Josh took the square root of a positive number, tripled that number, and then added 8, he got the number 20. But instead of taking the square root of the number, he should have squared it. If Josh had squared the number, what would his result be?

35. The sum of 10 consecutive integers, starting with 11, equals the sum of 5 consecutive integers, starting with what number?

36. A room is one-half full of people. After 20 people leave, the room is one-third full. How many people would be in the room if it were full?

37. A cell phone battery is at 40% and will run out in 36 minutes. How long (in minutes) will the battery last if it is at 90%?

38. My age is 9 less than 2 times my sister's age. In 4 years, the sum of our ages will be twice my brother's current age, which is 5 more than my sister's current age. What is my age?

39. The sum of 9 consecutive integers is 90. What is the largest of these integers?

40. If $x^2 - y^2 = 96$ and $x + y = 16$, then what is $x - y$?

41. Michelle is taking a geometry class in which an average of 85 or above on tests is needed to pass. Her scores so far have been 98, 86, 87, and 64. What is the lowest score she can get on her fifth test and still pass?

42. It takes 5 men to build 3 cars in 8 days. How many days will it take 2 men to build 6 cars?

43. A bike and a car are 210 miles apart. The bike is going at 10 miles per hour in the direction of the car and the car is driving at 60 miles per hour in the direction of the bike. How many hours will it take for the two vehicles to reach each other?

44. A function $f(x)$ is defined by $f(x) = f(x - 1) + f(x - 2)$. If $f(1) = 2$ and $f(5) = 25$, find $f(4)$.

45. If a driver drives five miles at 50 miles per hour, how fast should he drive for the next 3 miles so that his average speed for the whole trip is 60 miles per hour?

46. Bob has taken 6 tests so far in his math class, each worth a total of 100 points. His grade in math class depends solely on his test scores, and is currently exactly 88%. He is going to take the final next week, and wants to know if he can bring his grade up to 90%. The final is worth 150 points. What grade in percent must Bob get on his final to bring his grade up to 90%?

47. How many ways can $5 be formed with quarters and pennies?

48. If $x^2 + y^2 = 12$ where x and y are real numbers, what is the maximum value of xy?

49. Adam says that his favorite number is a positive number x such that $x = \cfrac{1}{x+\cfrac{1}{x+\cfrac{1}{x+\cdots}}}$. Compute Adam's favorite number.

50. If the answer to this problem is x, what is $x^2 + 3x + 1$?

51. At the beginning of a trip, the mileage odometer read $56,200$ miles. The driver filled the gas tank with 6 gallons of gasoline. During the trip, the driver filled his tank again with 12 gallons of gasoline when the odometer read $56,560$. At the end of the trip, the driver filled his tank again with 20 gallons of gasoline. The odometer read $57,060$. To the nearest tenth, what was the car's average miles-per-gallon for the entire trip?

52. Michelle wants to go to Pakistan to visit her dog. She goes online to buy tickets, but her computer has a mysterious virus that turns numbers into algebraic expressions. Pakistan

Airlines shows that her ticket costs $\frac{x^2-x-6}{x+2}$ dollars where $x = 1340$. How much money does her ticket cost (in dollars)?

53. What is the sum of the first 13 terms of the following arithmetic sequence: $30, 26, 22, 18, ...$?

54. Let x be an integer such that $x^2 - 8x + 11 < -4$. Compute x.

55. Jenny buys 10 cupcakes from Whipped Cream Cupcakes. Her friend Jessica buys 14 cupcakes at the store the next day. All of the cupcakes in the store cost $2. The cashier tells both of them that Whipped Cream Cupcakes offers a sale such that any purchase of 18 or more cupcakes gets 4 free cupcakes. How much could Jenny and Jessica have saved on their cupcakes if they had bought them together?

56. What is the measure in degrees of the acute angle formed by the hour and minute hands of a clock at 2 : 20 PM?

57. Compute the number of integers x that will make $\dfrac{2x^2 - 4x + 15}{x - 5}$ an integer.

58. If it takes 12 men 12 hours to build 12 robots, how much time does it take 6 men to build 6 robots?

59. In the book <u>Moles</u> by Souis Lachar, moles must dig holes shaped like a cube with length 6 feet. Three moles can dig 9 cubic feet in 9 minutes. How many moles does it take to finish a complete hole in 27 minutes?

60. If a woodchuck chucks 10 chucks of wood in 3 hours, how many chucks of wood can 3 woodchucks chuck in 1 hour?

61. Compute the value of $1 + \cfrac{1}{2 + \cfrac{1}{1 + \cfrac{1}{2 + \cfrac{1}{1 + \cdots}}}}$.

62. If the cost of an apple and 2 oranges is 60 cents, the cost of an orange and 2 peaches is 36, the cost of a peach and 2 apricots is 120, and the cost of an apricot and 2 apples is 96, what is the cost of buying one of each (in cents)?

63. Find the sum of all perfect cubes that are less than 250.

64. Let $a \wr b = \dfrac{a|b-a|}{b}$ and $a \dagger b = a \wr b + b \wr a$. What is $(4\dagger 6)\wr(6\dagger 4)$?

65. A movie theater sells adult tickets for 6 dollars and children tickets for 4 dollars. 300 people attended one showing, and the theater made \$1560. How many adults watched the movie?

66. Let $a \triangle b = \frac{a+b}{ab}$ and $a \angle b = \frac{b-a}{ab}$. What is $(123\triangle 456) + (41\angle 456)$? Express your answer as a common fraction.

67. In 6 years, I will be a third as old as my father. In 24 years, I will be half as old as him. How old am I?

Challenge Problems

1. For $x > 0$, given that $x^2 + 1/x^2 = 7$, compute $x + 1/x$.

2. Susan left her house at around 8:15 PM for a walk. When she came back, she realized that on her mechanical clock, the two pointers overlapped. She hasn't been out for more than an hour. When did she come back? Round your answer to the nearest minute. Express your answer in the form Hour : Minute.

3. Pipes A and B flow into a 1,000 liter tank while pipes C and D flow out. The tank starts out empty. Pipes A and B are opened, while C and D are closed, and the tank takes 4 hours to fill up. A and B are then closed, while C and D are opened, and the tank takes 6 hours to empty. Pipes A and B are opened, C is left open, and D is closed. The tank takes 8 hours to fill up. If pipes A, B, and C are closed, and D is opened, how long will it take for the tank to fill?

4. $f(x) = x^3 + ax^2 + bx + c$ is a cubic. If $f(-11) = f(-2) = f(2) = 48$, find an expression for $f(x)$.

5. If $x + \frac{1}{x} = a$ and $x^2 + \frac{1}{x^2} = a$, where x and a are real numbers, what are the possible values for a?

6. At sunrise, two people start to walk towards each other at constant, but different, speeds. One starts from town A and goes towards town B, while the other starts from town B and goes towards town A, following the same path in the opposite direction. At noon, they pass each other. The first person reaches town B at 4:00 p.m., while the second person reaches town A at 9:00 p.m. When was sunrise that day?

7. Paul and Rick are both painting rooms. Paul can paint a room in 3 hours, but if they work together, they can paint one room in 72 minutes. How many days would it take them to paint 43 rooms working together if Paul works for 8 hours a day and Rick works for 9 hours a day? Assume that the hours they spend working do not include travel time from one room to the next.

8. Compute: $\frac{1}{2\times4} + \frac{1}{4\times6} + \frac{1}{6\times8} + \cdots$

9. What is the minimum value of the expression $x^2 + 8y + z^2 + 25 - 6x + 15 + y^2 - 12z$? What is the value of xyz when this expression is at its minimum value?

One classic example of near perfect symmetry in nature is the butterfly. Each species of butterfly has a unique type of pattern on its wings. Sometimes, the spots are used to scare off predators, and other times, these patterns are used to distract predators from the vital body parts such as the head.

Chapter 2

Geometry

Warm-up Problems

1. What is the area of a circle whose circumference is equal to 12π? Express your answer in terms of π.

2. If right triangle ABC has $AB = 34$ and $BC = 16$, what is the maximum possible side length of AC?

3. Parallelogram ABCD has $\angle ABC = 50°$. The angle bisector of $\angle BCD$ meets side AB at a point E between A and B. What is the measure of $\angle CEA$?

4. Compute the distance (in units) between the points $(2, -1)$ and $(8, 7)$.

5. The side lengths of a triangle are 9 inches, 12 inches, and 15 inches. In square inches, what is the area of the triangle?

6. A 40-degree sector is cut out of a circle with radius 90 feet. What is the perimeter of the sector (in feet)? Express your answer in terms of π.

7. Johnny wants to plant flowers in a 3 feet by 3 feet square garden and place a stone walkway 3 feet wide around the garden (The outer edges of the walkway also form a square). If the flowers cost $4 per square foot and the walkway costs $6 per square foot, how much will Johnny have to pay for his garden and walkway (in dollars)?

8. Arthur wants to build a moat around his castle. The castle sits on a circular plot of land of radius 50 meters. If the moat is to be 10 meters wide all around, what area will the moat cover (in square meters)? Express your answer in terms of π.

9. Mr. Bob needs to paint a wall. A bucket of paint can cover 25 square meters. If Mr. Bob's wall is 15 meters wide and

9 meters tall, what is the minimum number of buckets Mr. Bob will need to paint the wall?

10. Tom has a 12 feet by 8 feet by 8 feet box. What is the maximum number of iron blocks with dimensions 2 feet by 2 feet by 4 feet that can fit inside Tom's box?

Introductory Examples

1. A circle with radius r inches has a circumference of at least 10 inches. Compute the smallest possible integer value of r that satisfies the condition.

   Solution: For a given radius r, we know that the circumference of a circle with radius r is $2\pi r$. We have $2\pi r \geq 10$, so $r \geq \dfrac{5}{\pi}$. Using the approximation $\pi \approx 3.14$, we get that $r \geq 1.59$. Since r must be an integer, the smallest possible radius is $\boxed{2}$.

2. How many triangular faces does a pyramid with 10 edges have?

   Solution: Let's first consider a pyramid with a quadrilateral as a base. There will be four triangular faces that surround the base (one on each side) and meet at a vertex. The sides of the quadrilateral account for four edges. There are also four edges that start from the vertex and go to each of the four vertices of the quadrilateral. Hence, for a pyramid with a 4-sided polygon as a base, we have 8 edges and 4 triangular faces.

   Building off of this, we see that a n-gonal pyramid (a pyramid with an n sided polygon as a base) has $2n$ edges and n triangular faces. Thus, a pyramid with 10 edges must have a pentagon as a base, and have $\boxed{5}$ triangular faces.

3. A point P is chosen inside square $ABCD$. Given that $\angle APB = x$, $\angle BPC = 2x - 30$, $\angle CPD = 3x$, and $\angle APD = 60$, what is the value of x?

   Solution: We know that the sum of all the angles around a point is 360. Then

   $$\angle APB + \angle BPC + \angle CPD + \angle APD = 360$$

$$x + 2x - 30 + 3x + 60 = 360$$

$$6x = 330$$

$$x = \boxed{55}$$

4. A 5 inch by 7 inch rectangle is cut out of a corner of a 16 inch by 19 inch rectangle. In inches, what is the perimeter of the remaining figure?

Solution: Interestingly, the perimeter does not change when the rectangle is cut out of the corner. See if you can prove this. So the perimeter is $2(16 + 19) = \boxed{70}$ inches.

5. The measure of an interior angle of a regular polygon is 171 degrees. How many sides does this polygon have?

Solution: In a polygon with n sides, the sum of all the interior angles is $180(n-2)$. In a regular polygon, the interior angles are equal, so we must have

$$\frac{180(n-2)}{n} = 171$$

$$180n - 360 = 171n$$

Solving the above equation, we get $9n = 360$, so the polygon has $\boxed{40}$ sides.

6. Let A be the point $(3,9)$ on the xy-plane. A is reflected across the y-axis to point A'. Then A' is reflected across the x-axis to point A''. Finally, A'' is reflected across the line $x = 10$. What is the sum of the coordinates of the image of A''?

Solution: We go through the transformations step by step, calculating the coordinates of the image of A for each step.

After reflecting A across the y-axis, the coordinates become $(-3,9)$. Reflecting across the x-axis produces $A''(-3,-9)$. Finally, reflecting across the line $x = 10$ gives us $(23,-9)$, so the sum of the coordinates is $23 + (-9) = \boxed{14}$.

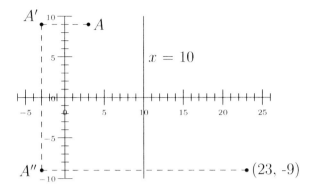

$x = 10$

$(23, -9)$

7. Tommy the Train is going at a rate of 60 miles per hour. From Town A, it goes north for 3000 miles, and then east for 4000 miles to reach town B. If Tommy the Train had gone directly from Town A to Town B in a straight line, how many minutes less would it take?

Solution: If Tommy the Train goes at 60 mph, he travels one mile a minute. Using his original path, Tommy travels $3000 + 4000 = 7000$ miles, so it takes him 7000 minutes. If he had gone in a straight line, Tommy would travel $\sqrt{3000^2 + 4000^2} = 5000$ miles in 5000 minutes. Thus, he travels for $7000 - 5000 = \boxed{2000}$ less minutes by traveling in a straight line.

8. Points $(1, 1)$ and $(5, 9)$ are opposite vertices of a square. What is the sum of the x-coordinates of the other two vertices?

Solution 1: The center of the square is the midpoint of the line connecting two opposite vertices of the square, so the center of the square has coordinates $(3, 5)$. Then the average of the other x-coordinates is equal to 3, so the sum of the x-coordinates is equal to $\boxed{6}$.

Solution 2: Let A be point $(1, 1)$ and B be point $(5, 9)$. Since A and B are opposite corners, the midpoint of segment AB is the center, O, of the square. We can compute the coordinates of O: $\left(\dfrac{1+5}{2}, \dfrac{1+9}{2}\right) = (3, 5)$.

Consider the right triangle drawn as shown with AO as the hypotenuse. Notice that when we rotate the triangle 90 degrees clockwise, the image of A is a vertex of the square. Similarly, when we rotate the triangle 90 degrees counterclockwise, we get another vertex. Using these right triangles, we can find the x-coordinates of the other two vertices. They are -1 and 7, so their sum is $\boxed{6}$.

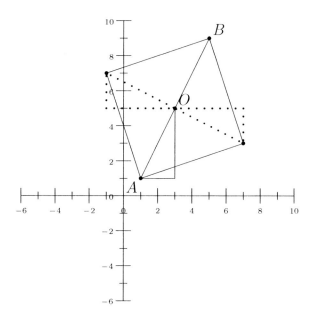

9. Triangle ABC is isosceles with $AB = AC$. If the angle bisector of B touches AC at D and $AD = BD$, compute the measure of angle BAC.

Solution: Let $\angle BAC = x$. Since $AD = BD$, $\triangle ABD$ is isosceles, so $\angle ABD = x$. However, BD is the angle bisector of

40

$\angle ABC$, so $\angle CBD = x$. This means that $\angle ABC = \angle ACB = 2x$.

From $\triangle ABC$, we can write that $x + 2x + 2x = 180$, or $5x = 180$. Dividing both sides of the equation by 5, we get $x = \boxed{36}$ degrees.

10. PS and RT are angle bisectors of triangle PQR (with S on QR and T on PQ) and $\angle PTR = 110$ degrees. Given $\angle RPS = 20$ degrees, compute the value of $\angle PQR$.

Solution: We start from what we have and work our way to $\angle PQR$. From $\angle RPS = 20$, we deduce that $\angle RPQ = 40$ because PS bisects $\angle RPQ$. Then looking at triangle RPT, we can compute $\angle PRT$:

$$\angle PRT = 180 - \angle RPQ - \angle PTR = 180 - 40 - 110 = 30$$

But RT bisects $\angle PRQ$, so $\angle PRQ = 2\angle PRT = 60$. Finally, looking at the angles of triangle PQR, we have $\angle PQR = 180 - \angle RPQ - \angle PRQ = 180 - 40 - 60 = \boxed{80}$ degrees.

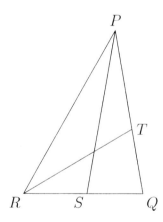

11. In triangle ABC, $AB = 2$, $\angle BAC = 105°$, and $\angle ABC = 45°$. What is the perimeter of $\triangle ABC$? Express your answer in simplest radical form.

Solution: From the two angle conditions, we see that $\angle ACB = 30$ degrees. We have a 30 degree angle and a 45 degree angle, so we should try to construct a 30-60-90 triangle and 45-45-90 triangle. To do this, we drop a perpendicular from A to BC and let it intersect BC at D:

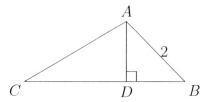

Now, $\triangle ABD$ is a 45-45-90 triangle and $\triangle ACD$ is a 30-60-90 triangle. It follows that $AD = BD = \sqrt{2}$, and $AC = 2\sqrt{2}$ and $CD = \sqrt{6}$. The perimeter of $\triangle ABC$ is then

$$2 + \sqrt{2} + \sqrt{6} + 2\sqrt{2} = \boxed{2 + 3\sqrt{2} + \sqrt{6}}$$

12. The circumcircle of triangle ABC has center O. Point D is on ray BC past C such that $CD = 9$ and $AD = 12$. If $BC = 7$, compute the value of $\angle DAO$.

   Solution: We are given a lot of lengths and secants AD and BD, but no angle measurements. This makes us think that $\angle DAO$ may be some special angle. Since we have secants, we can apply Power of a Point. The power of D is $DC \cdot DB = 9(9 + 7) = 144$. But $144 = 12^2 = DA^2$! This means that AD is tangent to the circumcircle of ABC, so $\angle DAO = \boxed{90}$ degrees.

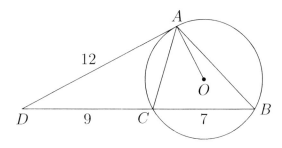

13. $ABCD$ is a rectangle such that $BC = x$, $CD = x\sqrt{3}$, and $\angle CAD = x$ degrees. Compute the value of AC.

Solution: Since $ABCD$ is a rectangle, $AD = BC = x$. Combining this with $\angle ADC = 90$ and $CD = x\sqrt{3}$, we find that triangle ACD is a 30-60-90 triangle. This implies that $x = \angle CAD = 60$ degrees! As a result, $AC = 2AD = 2x = \boxed{120}$.

14. The centers of two circles with radius 6 units and 13 units are 25 units apart. Compute the length of the common external tangent (in units).

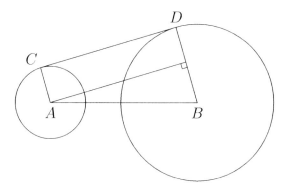

Solution: To compute the length CD, we form a right triangle as shown. The longer leg of the right triangle is equal to CD, and the length of the shoter leg is equal to $13 - 6 = 7$. Also, $AB = 25$, so by the Pythagorean Theorem, $CD = \sqrt{25^2 - 7^2} = \boxed{24}$ units.

15. A triangle has side lengths 6, 8, and x, where x is an integer. Compute the sum of all possible values of x.

Solution: The side lengths of a triangle must satisfy the Triangle Inequality. If x is the longest side, we have $8+6 > x$, so $14 > x$. If x is not the longest side, $6 + x > 8$, so $x > 2$. Combining our two inequalities, we have $2 < x < 14$, so x

must be an integer between 3 and 13 inclusive. The sum of all of these values is $\frac{(3+13)\cdot 11}{2} = \boxed{88}$.

16. Noakim Joah and Gaj Tibson are sweeping the floor. The floor is a rectangle with a semicircle on top of the longer side. The diagonal of the rectangle is 5000 centimeters, and one side is 30 meters. What is the area of the floor that the two have to sweep in square meters? Express your answer in terms of π.

Solution: The diagonal of the rectangle and the two sides form a right triangle. Using the Pythagorean Theorem, we find that the longer side of the rectangle is $\sqrt{50^2 - 30^2} = 40$ meters long. Then the area of the rectangle is $30 \times 40 = 1200$ square meters. Moreover, the radius of the semicircle is $\frac{40}{2} = 20$ meters, so the area of the semicircle is $\frac{20^2\pi}{2} = 200\pi$ square meters. Hence, the area that the two have to sweep is $\boxed{1200 + 200\pi}$.

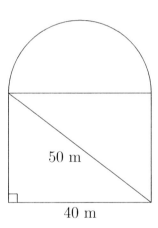

50 m

40 m

17. Median AD is drawn in $\triangle ABC$ with D on BC. Let H be the point on BC such that $AH \perp BC$. If $\angle HAC = 30$ degrees, $AD = 13$, and $HD = 5$, what is the area of $\triangle ABC$? Express your answer in simplest radical form.

44

Solution: Since ADH is a right triangle, we can use the Pythagorean Theorem to evaluate AH. We have $AH^2 + 5^2 = 13^2$, so $AH = 12$. Next, since $\angle HAC = 30$, triangle AHC is a 30-60-90 triangle, implying that $CH = \frac{AH}{\sqrt{3}} = 4\sqrt{3}$. Then $CD = CH + HD = 4\sqrt{3} + 5$.

Therefore, the area of ABC can be expressed as

$$\frac{BC \cdot AH}{2} = CD \cdot AH = (4\sqrt{3} + 5)12 = \boxed{60 + 48\sqrt{3}}$$

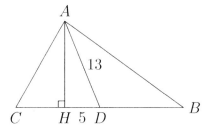

18. $ABCD$ is a trapezoid with bases $AB = 4$ and $CD = 10$. If $\angle BCD = 30°$ and $\angle ADC = 90°$, what is the area of $ABCD$? Express your answer in simplest radical form.

    Solution: The area of a trapezoid can be found by adding up the bases and multiplying the sum with half of the height. We have the bases already; we just need to find the height. We drop the perpendicular from B to CD. Let the perpendicular intersect CD at E. Since $\angle BCD = 30$, it follows that $\triangle BCE$ is a 30-60-90 triangle. We have $CE = CD - AB = 10 - 4 = 6$, so $BE = \frac{6}{\sqrt{3}} = 2\sqrt{3}$. Consequently, the area of $ABCD$ is $\frac{(4 + 10) \cdot 2\sqrt{3}}{2} = \boxed{14\sqrt{3}}$.

19. What is the inradius of a triangle with side lengths 5, 6, and 7? Express your answer as a fraction in simplest radical form.

    Solution: We can use area to relate the inradius with the side lengths of a triangle. We can calculate the area of a triangle given the three sides using Heron's Formula: Area $= \sqrt{s(s-a)(s-b)(s-c)}$, where $s = \dfrac{a+b+c}{2}$ and a, b, c are the lengths of the sides. In addition, the area of the triangle can also be expressed as rs, where r is the inradius. Therefore, we have $s = \dfrac{5+6+7}{2} = 9$, so

$$\sqrt{(9)(9-5)(9-6)(9-7)} = \text{Area} = 9r$$

$$\sqrt{(9)(4)(3)(2)} = 9r$$

$$6\sqrt{6} = 9r$$

implying that $r = \dfrac{6\sqrt{6}}{9} = \boxed{\dfrac{2\sqrt{6}}{3}}$.

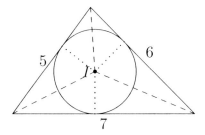

20. Let v be the number of vertices in a rectangular prism, e be the number of edges, and f be the number of faces. What is the value of $f + v - e$?

Solution: A rectangular prism contains 6 faces, and has 8 vertices connected by 12 edges. The answer is then $6+8-12 = \boxed{2}$.

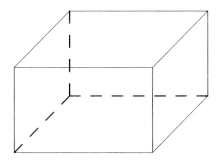

Note: For any polyhedron with f faces, v vertices, and e edges, the condition $f + v - e = 2$ is always satisfied. See if you can prove this.

Difficult Problems

1. Triangle ABC is inscribed in a circle. Let D be the point on arc AC such that $AD = CD$ and B and D are on opposite sides of AC. Let BD intersect AC at E. If $AB = 7$, $BC = 10$, and $AC = 13$, compute the length of AE. Express your answer as a common fraction.

   Solution: To compute AE, we have to find out more about point E. We know that E is the intersection of BD and AC, and D is chosen on the circle such that $AD = CD$. This implies that $\angle ABD = \angle CBD$, so BD is the angle bisector of $\angle ABC$. Therefore, E is the point where the angle bisector of $\angle ABC$ intersects AC.

   By the Angle Bisector Theorem,

$$\frac{BA}{AE} = \frac{BC}{CE} \implies \frac{AE}{CE} = \frac{BA}{BC} = \frac{7}{10}$$

   This means that AE and CE are in a $7 : 10$ ratio. Then we can let $AE = 7x$ and $CE = 10x$, where x is some positive number. Since $AC = 13$, $7x + 10x = 13 \implies x = \dfrac{13}{17}$. As a result, $AE = \dfrac{7 \times 13}{17} = \boxed{\dfrac{91}{17}}$.

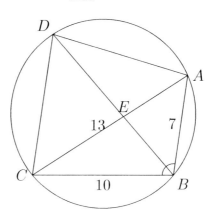

2. Let ABC be a triangle with $AB = 15$ and $AC = 21$. Let D be the point on BC such that AD bisects $\angle BAC$, and let F be the midpoint of AB. Let X be the point on line AD such that FX is perpendicular to AD. Let Y be the point on line AD such that CY is perpendicular to AD. What is the value of $\frac{AX}{AY}$? Express your answer as a common fraction.

Solution: We want to find the ratio between AX and AY. The first thing we can note is that both AX and AY are a part of AD and are each a part of two triangles AFX and AYC respectively. Maybe we can relate the two triangles?

From $FX \perp AD$ and $CY \perp AD$, we can say that $\angle AXF = 90 = \angle AYC$. However, $\angle FAX = \angle CAY$ because AD bisects $\angle BAC$. Therefore, $\triangle AFX \sim \triangle ACY$ by AA similarity. We then have

$$\frac{AX}{AY} = \frac{AF}{AC}$$

We know that F is the midpoint of AB, so $AF = \frac{15}{2}$. Hence,

$$\frac{AX}{AY} = \frac{\frac{15}{2}}{21} = \boxed{\frac{5}{14}}$$

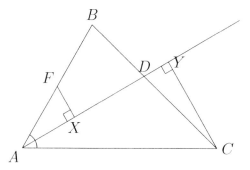

3. Let I be the center of the incircle of triangle ABC. The incircle is tangent to AB at M and AC at N. Let line AI intersect BC at D. Let P be a point on minor arc MN ($P \neq M$ and $P \neq N$). Let the line tangent to the incircle at P intersect AB at X and AC at Y. If $BI = 6$, $DI = 3$,

and $\angle ABC = 60$, compute the perimeter of triangle AXY. Express your answer in simplest radical form.

Solution: To find the perimeter, we need to relate A, X, and Y. Unfortunately, none of the given lengths and angle measurements tell us anything about those three points. However, looking at triangle BID, we note that $BI = 2DI$. Since I is the incenter of ABC, BI bisects $\angle ABC$, so $\angle IBD = \frac{60}{2} = 30$ degrees. This implies that triangle BID is a 30-60-90 triangle!

Hence, $\angle IDB = 90$, so AD is an altitude. However, AD also goes through I, so AD is an angle bisector as well. This makes ABC an isosceles triangle. Combining this with $\angle ABC = 60$, we find that ABC is in fact an equilateral triangle.

Since both XP and XM are tangents to the incircle, $XP = XM$. By similar reasoning, $YP = YN$. Then the perimeter of AXY is $AX + XP + YP + AY = AM + AN$. Because ABC is equilateral, $AM = AN = BD$, so the perimeter can be expressed as $2BD$. Finally, using the 30-60-90 triangle BID, we get $BD = 3\sqrt{3}$, so the perimeter of AXY is $\boxed{6\sqrt{3}}$.

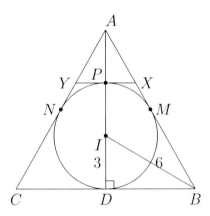

4. Let ABC be a triangle with $AB = 15$, $AC = 13$, and $BC = 14$. Let AD be the angle bisector of $\angle BAC$ with D on BC. Let H be the point on BC such that $\angle AHB = 90$. Compute

the ratio of the area of triangle AHC to the area of triangle ABD. Express your answer as a common fraction.

Solution: We first note that both AHC and ABD share the same height, AH. Then the ratio between the areas is the same as the ratio between the bases, so we want to find $CH : BD$.

Since AD is the angle bisector of $\angle BAC$, we can compute BD using the Angle Bisector Theorem:

$$\frac{AB}{BD} = \frac{AC}{CD} \implies \frac{15}{BD} = \frac{13}{14 - BD}$$

Cross multiplying, we have $15 \cdot 14 - 15 \cdot BD = 13 \cdot BD \implies 15 \cdot 14 = 28 \cdot BD$, so $BD = \frac{15}{2}$.

To compute CH, we can use the Pythagorean Theorem on AHC and AHB. Note that

$$AC^2 - CH^2 = AH^2 = AB^2 - BH^2$$

$$13^2 - CH^2 = 15^2 - (14 - CH)^2$$

Expanding and rearranging, we have

$$140 = 28 \cdot CH$$

so $CH = 5$. Therefore, $\dfrac{CH}{BD} = \dfrac{5}{\frac{15}{2}} = \boxed{\dfrac{2}{3}}$.

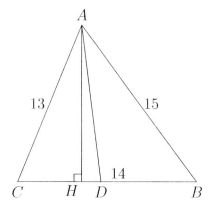

5. $EBCD$ is a parallelogram and ABC is a triangle such that E lies on segment AB. AC intersects DE at M, and M is the midpoint of DE. If the area of $\triangle ABC$ is 24, what is the sum of the areas of CDM and AEM?

Solution: We need to find the sum of the areas of CDM and AEM, so we first can try to find a relationship between the two triangles. First, we note that $AE \| CD$ due to parallelogram $EBCD$. This gives us a few angle equalities, including $\angle AEM = \angle CDM$ and $\angle EAM = \angle DCM$. It follows that $\triangle AEM \sim \triangle CDM$ by AA similarity. But we also know that $DM = EM$ (M is midpoint of DE), so the two triangles are actually congruent!

Unfortunately, that alone won't solve the problem. We still haven't used the fact that the area of ABC is 24, so we should try to relate ABC to AEM or CDM. Since $BC \| DE$, we have $\angle AEM = \angle ABC$ and $\angle AME = \angle ACB$, meaning that $\triangle AEM \sim \triangle ABC$. Furthermore, since $\triangle AEM \cong \triangle CDM$, $AM = CM$. This means that the ratio between the sides of AEM and ABC is $1 : 2$, so the area ratio is $1 : 4$. Thus, the area of AEM is 6, so the sum of the areas of AEM and CDM is $6 + 6 = \boxed{12}$.

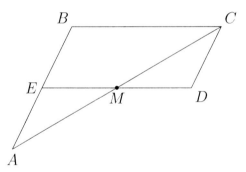

6. Let A and B be the points $(3, 4)$ and $(5, 12)$, respectively, on the Cartesian plane. A point C is chosen on the y axis. Compute the minimum possible value of $AC + BC$. Express your answer in simplest radical form.

Solution: We are asked to find the shortest distance from A to C to B. We know that given two points X and Y, the shortest distance from X to Y is just the straight line from X to Y. However, in this case, we are dealing with three points, and they aren't collinear...

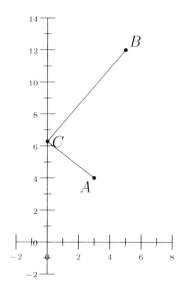

However, we can make them collinear by reflecting B across the y-axis to get B', located at $(-5, 12)$. Observe that $BC = B'C$, so $AC + BC = AC + B'C$. Now, the optimal point C would be the point where line segment AB' intersects the y-axis! As a result,

$$AC + BC = \sqrt{(3+5)^2 + (12-4)^2} = \sqrt{8^2 + 8^2} = \boxed{8\sqrt{2}}$$

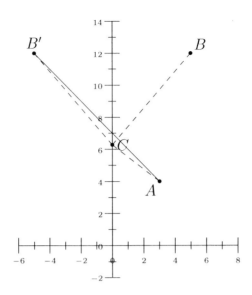

7. Square $ABCD$ has a side length of 1. Minor arc $\overarc{BD}$ of a circle with radius 1 centered at C is drawn along with diagonal BD. Let CE be the altitude of $\triangle BCD$ drawn from C. Line CE is extended to intersect minor arc $\overarc{BD}$ at F. Compute the length of $2EF$. Express your answer in simplest radical form.

Solution: Rather than directly finding EF, we can compute the value of CF and CE, because their difference is EF. Since F is on the quarter circle, $CF = 1$. Furthermore, CE is the altitude of a 45-45-90 triangle BCD. This means that CED is also a 45-45-90 triangle. Since $BD = CD\sqrt{2} = \sqrt{2}$, we have $CE = DE = \frac{BD}{2} = \frac{\sqrt{2}}{2}$. Consequently,

$$2EF = 2(CF - CE) = 2\left(1 - \frac{\sqrt{2}}{2}\right) = \boxed{2 - \sqrt{2}}$$

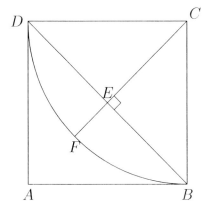

8. $ABCD$ is a quadrilateral with $\angle DAB = \angle ABC = 135$ and $\angle BCD = 30$. If $AB = \sqrt{2}$ and $AD = 2$, compute CD.

Solution: We are given three interior angles of $ABCD$, so we can easily calculate $\angle ADC = 360 - \angle DAB - \angle ABC - \angle BCD = 60$. So we have $\angle ADC = 60$ and $\angle BCD = 30$. This reminds of 30-60-90 triangles, but we don't have any in our diagram. Fortunately, we can make one by extending AD past A and BC past B! If we let lines AD and BC intersect at E, we have a 30-60-90 triangle with CDE. Even better, since $\angle DAB = \angle ABC = 135$, we have $\angle EAB = \angle EBA = 45$, so ABE is a 45-45-90 triangle!

From $AB = \sqrt{2}$, we see that $AE = 1$. Then $DE = AE + AD = 1 + 2 = 3$, so $CD = 2DE = \boxed{6}$.

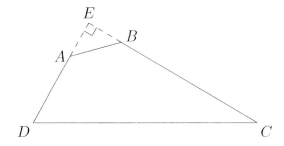

9. An ant is on one vertex of a unit cube. It wants to go to the opposite corner by walking along the surface of the cube. What is the minimum distance the ant must walk? Express your answer as a radical in simplest form.

Solution: Because the ant can only walk on the surface, instead of considering a cube, we can consider the net of the cube. Thus, we unfold the cube, and now we just need to find the distance between the two points as shown. By the Pythagorean Theorem, the minimum distance the ant walks is $\sqrt{1^2 + 2^2} = \boxed{\sqrt{5}}$.

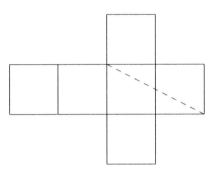

10. Barney has a cylinder of circumference 6 and height 24. He wraps a string around the lateral surface of the cylinder starting from the bottom. The string goes around the cylinder exactly three times before reaching the top of the cylinder. What is the length of the string?

Solution: Since we are dealing with an object traveling across the surface of a three-dimensional figure, it is a good idea to look at the net of the figure. Here, we "open" the cylinder and look at the net of the lateral surface:

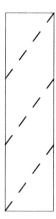

Since the string wraps around three times, the height is divided into three sections of length 8. The width (shorter side) of the rectangle is 6 (the circumference), so using the Pythagorean Theorem, we see that the dashed lines have length 10. As a result, the length of the whole string is $\boxed{30}$.

Additional Problems

1. A square and an equilateral triangle share a side. What is the ratio of the area of the triangle to the area of the square?

2. Circle ω with center O has radius 6. Consider two points P and Q on ω such that $OP = PQ$. Compute the area of triangle OPQ.

3. A rectangle has its length and width increased by 20%. By how many percent will the area increase?

4. A square and a triangle have equal perimeters. The lengths of the three sides of the triangle are 6.2 cm, 8.3 cm and 9.5 cm. Compute the area of the square in square centimeters.

5. What is the area (in square units) of a triangle with coordinates $(1, 2), (3, 0)$, and $(3, 3)$?

6. A triangle has two side lengths 5 and 7. If the third side is an integer, what is the sum of all possible lengths of the third side?

7. Michelle sees a penny 3 feet from her shoes. If her hand is 4 feet above her shoes, how far, in feet, is her hand from the penny?

8. Farmer Bob needs to fence in an area of his farm to grow yummy strawberries. Unfortunately, he is on a short budget and can only buy 50 feet of fence to enclose this strawberry patch. Compute the greatest area in which Farmer Bob can grow his strawberries, such that the entire area is fenced in.

9. How many different isosceles triangles have integer side lengths and perimeter 19?

10. Two concentric circles are drawn such that one circle's radius is 3 times the other. If the area between the two circles is 72π, what is the radius of the larger circle? (Two circles are concentric if they share the same center.)

11. The width of a rectangular prism is 3. Its length and height are both 5. What is its surface area?

12. If the lengths of two of the sides of an isosceles triangle are 6 and 12, then what is the perimeter of this triangle?

13. The length and width of a rectangle are doubled. What is the percent increase between the new area and the original area?

14. A man walks on a path around his house. Starting from his house, he first walks 30 feet east, then 40 feet north, then 60 feet west, then finally 80 feet south. How far, in feet, is he from his house?

15. I am trying to draw a right triangle with a hypotenuse length of 15 and a leg length of 4. How long will the other leg be?

16. A rectangle has a length of 19 less than two times the width. The diagonal of the rectangle is 13. What is the length of the rectangle?

17. Consider a right triangle ABC with $AB = 3, BC = 4$, and $CA = 5$. Consider points P, Q, and R on AB, BC, and CA, respectively such that $PBQR$ is a square. Compute the side length of that square.

18. Let $ABCDEFGH$ be a cube of side length 5 such that face $ABCD$ is the top face, E is directly below A, F directly below B, G directly below C, and H directly below D. Let S be a unit cube that also has A as a vertex and one of its faces is completely on face $ABCD$. If P is any vertex on S, compute the maximum value of PG^2.

19. Right triangle $\triangle ABC$ has $AB = 6$, $BC = 8$, and $CA = 10$. What is the length of the altitude from B to side AC? Express your answer as a common fraction.

20. Two circles of radius 3 do not intersect and are contained inside a square. Compute the minimum area of this square.

21. A cone sits on the table. It has a height of 30 and a radius of 8. A smaller cone with radius 4 is chopped off from the top (the cut is parallel to the base of the cone). What is the volume of the object that is left over? Express your answer in terms of π.

22. Isosceles right triangle ABC has hypotenuse AC. Square $ACDE$ shares an edge of length 1 with triangle ABC. If $ABCDE$ is a convex pentagon, what is the ratio of the perimeter of $ABCDE$ to its area?

23. Charlie is blowing up balloons for his birth-
day party. Each balloon is a perfect sphere
and has a diameter of 30 cm. After he blows
them up, each of the balloons leaks air at
a rate of 4π cm^3 of air every minute. How
many minutes does it take for the balloons
to shrink to a radius of 12 cm?

24. A regular hexagon is inscribed in a circle. Another regular
hexagon is circumscribed about the same circle. What is the
ratio of the area of the smaller hexagon to the area of the
larger hexagon?

25. Find the area of a triangle with side lengths 17, 25, and 28.

26. A rhombus is drawn such that one of its angles is 60°. If
the side length of the rhombus is 8, what is the length of the
longer diagonal?

27. The perpendicular bisectors of sides AB and AC of triangle
ABC meet at O. If $\angle COB = 139°$ and $\angle BOA = 97°$, what
is $\angle ABC$?

28. A lamp post and a nearby fire hydrant are
3 meters away from a wall. The lamp post
casts a 3.5 meter tall shadow on the wall.
The fire hydrant, which is 1 meter tall, casts
a 2 meter long shadow along the ground. In
meters, how tall is the lamp post?

29. A spherical balloon i=s being blown up. Its radius is increasing at a rate of $r = \sqrt{12t}$, where t is the time elapsed

in seconds and r is the radius in centimeters. What is the volume of the balloon after 3 seconds (in cubic centimeters)? Express your answer in terms of π.

30. An equilateral triangle is inscribed in a circle with a radius of 2 inches. In square inches, what is the area of this triangle? Express your answer in simplest radical form.

31. In square units, what is the maximum possible area of a rectangle with a perimeter of 10 units? Express your answer as a common fraction.

32. A regular octagon is made by cutting isosceles right triangles out of the corners of a 12 inch by 12 inch square. In inches, what is the perimeter of the octagon? Express your answer in simplest radical form.

33. The volume of a cone and cylinder are equal. If the heights of the two objects are equal, what is the ratio of the radius of the cone to the radius of the cylinder? Express your answer in simplest radical form.

34. A triangle with side lengths 12 units, 16 units, and 20 units is inscribed in a circle. In square units, what is the area of the circle? Express your answer in terms of π.

Challenge Problems

1. Billy Bob drew a quadrilateral $ABCD$ on the chalkboard. If $\angle BCD + \angle DAB = 180°$, $\angle CDB = 33°$, and $\angle ABC = 61°$, what is the measure of $\angle BCA$?

2. Triangle ABC is an equilateral triangle with a side length of 6. Semicircles with diameters AB, BC, and AC are drawn outside triangle ABC. Circle O circumscribes the new figure. The area of the region outside the 3 semicircles and triangle ABC but inside Circle O can be expressed in the form $\left(a\sqrt{b} - \frac{c}{d}\right)\pi - e\sqrt{f}$, where a, b, c, d, e, f are positive integers, c, and d are relatively prime, and b and f are not divisible by a perfect square of a prime. Compute $a + b + c + d + e + f$.

3. Let D be the point on segment BC of triangle ABC such that $\angle BAD = \angle CAD$. If $AB = 13$, $AC = 14$, and $BD = \frac{65}{9}$, compute the value of the inradius of $\triangle ABC$.

4. Anthony and Bob are walking along the figure below, a regular hexagon. They both start out at A and walk at the same speed. Anthony takes the path ABCDEFA, walking around the hexagon once. Bob takes the path ACDEFA, taking a shortcut from A to C. If Anthony takes 6 minutes to walk around the hexagon, how many minutes does it take Bob to walk his path? Express your answer in the form $a + \sqrt{b}$, where a and b are positive integers.

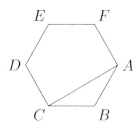

5. A unit cube has a right cylinder cut out of it with height $\frac{7}{13}$ and radius $\frac{1}{3}$. What is the surface area of the remaining object? Express your answer in the form $a + \frac{b}{c}\pi$.

One of nature's greatest phenomena is the snowflake. Snowflakes come in a large variety of shapes and sizes. In fact, no two snowflakes are alike! Despite their differences, all snowflakes have a basic hexagonal shape; the water molecules that form snowflakes naturally attract each other to combine in a symmetrical, 6-sided arrangement.

Chapter 3

Counting and Probability

Warm-up Problems

1. There are 61 families of turtles in Turtleville. Each family has 11 turtles, and each turtle has 3 shells. How many shells are there in Turtleville?

2. A cowboy plans to build a fence to enclose a square pasture. The perimeter of the plot is 96 feet, and he sets posts along the perimeter of the square, keeping adjacent posts 6 feet apart. How many posts will he use to fence the entire plot?

3. Abraham and Belinda are picking out sandwiches for lunch. They must choose a protein (egg, ham, or turkey), a vegetable (lettuce, pickles, or tomatoes), and a type of bread (whole grain, sourdough bread, or white bread). Belinda can only eat egg as a protein. Abraham does not eat tomatoes. However, they must order the same sandwich. How many different sandwiches can they order?

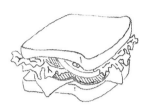

4. If the probability that you get a problem right on any test is 50%, and you take a test with 200 problems, how many questions on the test would you expect to get right?

5. Leonard and Sheldon went to lunch at the Chocolatecake Factory, and they each order 1 item from the menu. The menu has 10 items, and Sheldon refuses to eat what Leonard eats. In how many ways can they have a lunch that satisfies their requirements?

6. Batch rolls a regular 6-sided dice and flips 2 fair coins. What is the probability that he rolls a 3 on the die and flips two heads on the coins? Express your answer as a common fraction.

7. Canton Math Village is made up of a system that contains 3 leaders. Each leader trains 5 generals. Each general trains 1 apprentice. No general is under more than one leader, and no

apprentice is taught by more than one general. If everyone in the village is in this system, how many apprentices are there in Canton Math Village?

8. In how many ways can 4 be written as the sum of two or more (not necessarily distinct) positive integers? (Note that order does not matter, so $3 + 1$ is the same as $1 + 3$.)

9. Rancho San Joaquin Middle School has 1200 students. Each student takes 5 classes a day. Each teacher teaches 4 classes. Each class has 30 students and 1 teacher. How many teachers are there at Rancho San Joaquin Middle School?

10. There are 8 Martian letters in the Martian alphabet. A word in the Martian alphabet consists of 4 Martian letters. How many possible Martian words are there?

Introductory Examples

1. There are 355 students in Orange County High School, and each student takes Algebra II, Biology, or both. If 250 of the students are taking Algebra II, and 255 are taking Biology, how many students are taking only Biology?

   Solution: There are $250 + 255 = 505$ students taking Algebra II or Biology, but this double-counts those taking both classes. If we let x be the number of people taking both classes, we have $505 - x = 355$, so $x = 150$. As a result, the number of people taking only Biology is $255 - 150 = \boxed{105}$.

2. In *Ice Cream Town*, each of the 500 people like at least one of the three flavors: strawberry, vanilla, and chocolate. Exactly one-fifth of the people like strawberry ice cream, half of the people like vanilla, and 200 people like chocolate. Furthermore, 225 people like only vanilla, while only 10 people like both vanilla and strawberry. If 5 people like all three flavors, how many people like strawberry and chocolate but not vanilla?

   Solution: We can draw a Venn Diagram to illustrate the favorite ice cream types. First, we know that 5 people like all flavors, so we can fill in the center. If 10 people like both vanilla and strawberry, then 5 people like only vanilla and strawberry (we subtract off the people who like all three). In addition, 225 people like only vanilla, and since there are 250 (half or 500) people who like vanilla, $250 - 225 - 10 = 15$ people like only vanilla and chocolate.

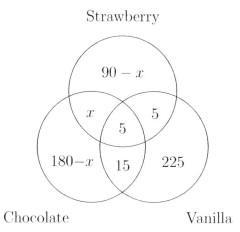

Now, we are left with three spaces to fill. Let x be the number of people who like only strawberry and chocolate. Looking at the strawberry circle and noting that $\frac{1}{5} \cdot 500 = 100$ people like strawberry, we have that $100 - 5 - 5 - x = 90 - x$ people like only strawberry. Similarly, $200 - x - 5 - 15 = 180 - x$ people like only chocolate.

Now, we add up all of the numbers in the Venn Diagram to solve for x:

$$90 - x + x + 5 + 5 + 225 + 15 + 180 - x = 500$$

$$520 - x = 500$$

$$x = \boxed{20}$$

3. Michael has a square pasture whose corners are located at $(0,0)$, $(0,20)$, $(20,20)$, and $(20,0)$. He wants to plant a tree at every lattice point inside or on his pasture, but he does not want to put a tree at any of his four corners. How many trees will Michael plant? (A lattice point is a point on the coordinate plane with integer value coordinates.)

Solution: We can first count how many lattice points are included inside or on Michael's pasture. Observe that there are 21 rows of lattice points (there are 21 points from $(0,$

0) to (0, 20), inclusive), and by similar reasoning, each row contains 21 lattice points. As a result, there are $21 \times 21 = 441$ lattice points total. Since the 4 corner lattice points will not contain trees, Michael will plant $441 - 4 = \boxed{437}$ trees total.

4. A yogurt shop has 4 different flavors, 6 different toppings, and 7 different cups. A combination consists of two flavors, two or three toppings, and one cup. How many different combinations can one make?

   Solution: We will build a yogurt and keep track of how many choices we have as we go. For our flavors, we have to choose 2 out of the 4 flavors, giving us $\binom{4}{2} = 6$ choices. Then we can choose either 2 or 3 toppings, so we have $\binom{6}{2} + \binom{6}{3} = 15 + 20 = 35$ combinations of toppings. Finally, we have to choose one cup, and there are 7 possible types of cups, so we have 7 choices there. Multiplying all of our choices together, we find that there are $6 \times 35 \times 7 = \boxed{1470}$ combinations.

5. How many four letter sequences can be made from the word MATHEMATICS if you cannot repeat letters? For example, TMCS and ATMI are valid sequences, but AHTT or SAIA are not.

   Solution: Since we cannot repeat letters, we are actually only dealing with 8 letters: M, A, T, H, E, I, C, S. We have to make a four-letter word with these letters, so there are 8 choices for the first letter in the word, 7 choices for the second, 6 choices for the third, and 5 choices for the last. Hence, there are $8 \times 7 \times 6 \times 5 = \boxed{1680}$ words total.

6. How many ways can the word $MATH$ be arranged so that A is not next to T?

   Solution: Without restrictions, there are $4! = 24$ ways to rearrange the letters. However, we must subtract the number of times A is next to T. If we treat AT as a unit, this gives $3! = 6$ ways. Similarly, if we treat TA as a unit, there are another 6 ways. Thus, our answer is $24 - 6 - 6 = \boxed{12}$.

7. In how many ways can the letters in $NEWYEAR$ be rearranged to form a seven-letter sequence?

Solution: The string $NEWYEAR$ contains seven letters, so there are $7! = 5040$ ways to arrange them. However, there are two E's and they are indistinguishable. This means that we have counted every string twice. Thus, we divide by $2! = 2$ to get $\frac{5040}{2} = \boxed{2520}$.

8. Six people are to be seated around a round table. Two of the people, Kate and Debbie, refuse to sit next to each other. How many distinct arrangements of the six people are there? Rotations of some arrangement are not distinct to one another.

Solution: To account for rotations, we can fix the seat of a person, say Kate, on the table. Then out of the remaining five seats, Debbie can only sit in three of them because the other two are next to Kate. After we pick Debbie's seat, the rest of the four seats can be filled in any way, giving us $4! = 24$ arrangements. In total, we have $3 \times 24 = \boxed{72}$ possible arrangements with Kate and Debbie not next to each other.

9. Define a binomial, $\binom{n}{r}$, as $\dfrac{n!}{r!(n-r)!}$, where $k! = (k)(k-1)(k-2)\cdots(2)(1)$ for all positive integers k.

Evaluate $\dfrac{\binom{10}{8}\binom{6}{2}}{\binom{7}{4}}$. Express your answer as a common fraction.

Solution: We evaluate each binomial separately. However, we don't multiply all of the factors together because we may cancel things out later.

$$\binom{10}{8} = \frac{10!}{8!2!} = \frac{10 \times 9}{2 \times 1} = 5 \times 9$$

$$\binom{6}{2} = \frac{6!}{2!4!} = \frac{6 \times 5}{2 \times 1} = 3 \times 5$$

$$\binom{7}{4} = \frac{7!}{4!3!} = \frac{7 \times 6 \times 5}{3 \times 2 \times 1} = 7 \times 5$$

71

Hence,

$$\frac{\binom{10}{8}\binom{6}{2}}{\binom{7}{4}} = \frac{5 \times 9 \times 3 \times 5}{7 \times 5} = \boxed{\frac{135}{7}}$$

10. There are 4 people at a party. Every person shakes every other person's hand once. How many total handshakes will take place?

    Solution: There are $\binom{4}{2} = 6$ ways to choose the two people who are shaking hands, so there are $\boxed{6}$ handshakes.

11. Paula the Painter has 7 pieces of classical music, 3 pop songs, and 1 rap song on her iPod. In how many ways can she choose 3 classical pieces and 2 songs from another genre?

    Solution: There are $\binom{7}{3} = 35$ ways to choose her three classical pieces. There are $3 + 1 = 4$ songs of a different genre, and Paula can choose two of them in $\binom{4}{2} = 6$ ways. As a result, there are $35 \times 6 = \boxed{210}$ ways in all.

12. Leah needs to travel to school from her home by traveling 4 blocks east and 5 blocks north. Due to traffic difficulties, she can only travel east and north. Compute the number of ways she can travel from her home to her school.

    Solution: In order to go to her school, she will need to travel one block east four times, and one block north five times. Thus, this is equivalent to the number of ways to order 5 N's and 4 E's, or $\binom{5+4}{4} = \boxed{126}$ ways.

13. In a room full of 61 people, what is the least number of people that must share a birth month?

    Solution: There are 12 months in a year, so by the Pigeonhole Principle, at least $\left\lceil \dfrac{61}{12} \right\rceil = \boxed{6}$ people must share the same birth month. (Note that if there were only five people sharing the same birth month, we would have at most $5 \times 12 = 60$ people, which is not enough.)

14. A pizza is cut into 16 equal wedge-shaped slices. 49 pepperonis are randomly sprinkled onto the pizza. Then one slice must have at least n pepperonis on it. Compute the largest possible value of n. Assume that pepperonis cannot land in between slices.

    Solution: By the Pigeonhole Principle, there must exist one slice with $\left\lceil \dfrac{49}{16} \right\rceil = 4$ slices of pepperoni. If all slices had at most three pepperoni slices, we would only have at most $16 \times 3 = 48$ pepperonis and $48 < 49$. On the other hand, there does not always have to be at least one slice with 5 pepperonis. For instance, if 15 of the slices had 3 and one had 4, that would give us a total of $15 \times 3 + 4 = 49$ pepperonis. Hence, the maximum value of n is $\boxed{4}$.

15. The pages of a book are numbered from 1 to 2014. How many digits are used when numbering the book? (Leading zeroes are not used in the numbering.)

    Solution: We can break this problem down into smaller cases based on the number of digits of each page number.

    Case 1: One-digit numbers. There are 9 one-digit numbers, and each one contributes one digit. Thus, 9 digits are used to print these numbers.

    Case 2: Two-digit numbers. There are $9 \times 10 = 90$ two-digit numbers. Each one contributes two digits, so $90 \times 2 = 180$ digits are used altogether.

    Case 3: Three-digit numbers. There are $9 \times 10 \times 10 = 900$ three-digit numbers. Each one contains three digits, so we print $900 \times 3 = 2700$ digits in this case.

    Case 4: Four-digit numbers. From 1000 to 2014, there are $2014 - 1000 + 1 = 1015$ four-digit numbers. Each one contributes four digits, giving us $1015 \times 4 = 4060$ digits altogether.

Adding up the results of the four cases, we see the answer is $9 + 180 + 2700 + 4060 = \boxed{6949}$.

16. King Arthur and his 10 knights are sitting around a circular table. A knight leaves his seat and inserts himself between two of the knights. What is the probability that he is sitting next to at least one of his original two neighbors? Express your answer a common fraction.

    Solution: After the knight leaves, there are 10 people around the table, so the knight has 10 spots to insert himself in. Of the 10 spots, 3 of them will cause him to sit next to one of his original neighbors. Thus, the probability he sits in one of those three spots is $\boxed{\dfrac{3}{10}}$.

17. The probability that a parachute opens in a testing facility is 0.7. 40% of the parachutes are from Company B, and 60% of the parachutes are from Company A. What is the probability that a randomly chosen parachute is a working parachute from Company A? Express your answer as a decimal rounded to the nearest hundredth.

    Solution: We first need to find the probability that the parachute we chose is from Company A. Since we know that of all the parachutes, 60% of them come from Company A, the probability a randomly chosen parachute is from Company A is 0.6. Then we multiply this decimal with the probability that the parachute works, which is 0.7. Thus, the probability our parachute is from Company A and that it opens is $0.6 \times 0.7 = \boxed{0.42}$.

18. A coin is flipped three times and a fair die is rolled once. What is the probability heads are flipped all three times and the die rolls a 2? Express your answer as a common fraction.

Solution: The probability that heads are flipped all three times is $\left(\dfrac{1}{2}\right)^3 = \dfrac{1}{8}$. The probability 2 shows up on the die is $\frac{1}{6}$. Hence, to find the probability both events occur, we multiply the separate probabilities to get $\dfrac{1}{8} \times \dfrac{1}{6} = \boxed{\dfrac{1}{48}}$.

19. A normal six sided dice is used in a children's game. Whenever the dice is rolled, a child places a sticker on the side facing up if there is no sticker on it already. What is the probability that after six rolls, all the faces will be covered by stickers? Express your answer as a common fraction.

Solution: We can calculate the final probability by considering the probabilities at each of the rolls, and then multiply them together. Since we need to place stickers on all six faces in six rolls, we must be placing a sticker at every roll.

For the first roll, none of the faces have a sticker yet, so it does not matter which side comes up. Therefore, the probability we place a sticker is $\frac{6}{6}$.

For the second roll, one of the faces has to have a sticker on it, so now we only have 5 faces left to choose from. Thus, the probability is $\frac{5}{6}$.

Similarly, the probability that we place a sticker on the third roll is $\frac{4}{6}$, and so on.

Multiplying all of these fractions together, we find that the final probability is

$$\frac{6}{6} \times \frac{5}{6} \times \frac{4}{6} \times \ldots \times \frac{1}{6} = \boxed{\dfrac{5}{324}}$$

20. Garry is trapped in a room and needs a key to get out. There are 14 dolls in the room, 8 with blue eyes and 6 with red eyes. Garry knows a doll with blue eyes is holding the key, but doesn't know which one. If Garry won't pick a doll that he knows does not hold the key, what is the probability that he guesses correctly on his second guess? Express your answer as a common fraction.

Solution: Garry knows that a blue-eyed doll carries the key, so he won't pick a red-eyed doll. Therefore, for him to miss on his first guess, he must pick one of the seven blue-eyed dolls, giving us a probability of $\frac{7}{8}$. On his second guess, note that Garry has only 7 dolls to choose from because he will not pick the doll he chose on the first guess (that doll cannot hold the key). Since he has to guess right on his second try, the probability is $\frac{1}{7}$. Consequently, the probability Garry picks the right doll on his second guess is $\frac{7}{8} \times \frac{1}{7} = \boxed{\frac{1}{8}}$.

21. In the game *Monster-Gold*, you encounter monsters that have gold. You have a one-third chance of surviving every time you meet a monster. If you survive an encounter with a monster, you take its gold (all monsters have a piece of gold). In addition, you begin with two lives, and lose one each time you don't survive the encounter with a monster. You need three pieces of gold to win, but you lose if you lose all of your lives. What is the probability that you win? Express your answer as a common fraction.

Solution: We can consider the cases in which we win. Given a string of games, denote W if we win and take the monster's gold, and L if we lose a life.

We need three pieces of gold, so one way is to not lose any lives and get three consecutive wins: WWW. On the other hand, we can lose one life at most before we lose, so the other winning arrangements are $(...)W$, where $(...)$ is a permutation of WWL. Note that there are 3 such permutations of $(...)W$.

The probability we achieve WWW is $\left(\frac{1}{3}\right)^3 = \frac{1}{27}$. The probability we get $(...)W$ is

$$3 \cdot \frac{1}{3} \cdot \frac{1}{3} \cdot \frac{2}{3} = \frac{2}{27}$$

Hence, adding the probabilities for the cases, we find that the probability we win is $\dfrac{1}{27} + \dfrac{2}{27} = \dfrac{3}{27} = \boxed{\dfrac{1}{9}}$.

22. There are 7 blue marbles and 3 red marbles in one bag, and there are 5 blue marbles and 5 red marbles in another bag. A marble is chosen from each bag. What is the probability that both of the marbles chosen are red? Express your answer as a common fraction.

Solution: There are 3 red marbles in the first bag, and there are $7+3 = 10$ marbles total in that bag. Thus, the probability a randomly chosen marble in that bag is red is $\frac{3}{10}$. Similarly, the probability a red marble is chosen in the second bag is $\frac{5}{5+5} = \frac{5}{10} = \frac{1}{2}$. It follows that the probability both marbles are red is equal to $\dfrac{3}{10} \times \dfrac{1}{2} = \boxed{\dfrac{3}{20}}$.

23. Box A contains two red books and one blue book. Box B contains three blue books and five red books. A box is chosen randomly and a book is picked from the chosen box. What is the probability a red book is chosen? Express your answer as a common fraction.

Solution: If Box A is chosen, the probability we choose a red book is $\frac{2}{3}$. On the other hand, if Box B is chosen, the probability we choose a red book is $\frac{5}{8}$. There is an equal chance of choosing either box, so our final probability is

$$\frac{1}{2} \cdot \frac{2}{3} + \frac{1}{2} \cdot \frac{5}{8}$$

$$= \frac{1}{3} + \frac{5}{16} = \boxed{\frac{31}{48}}$$

24. Brian rolls a six-sided die and a four-sided die. He then creates a two-digit number by using the number on the six-sided die as the tens digit, and the number on the four-sided die as the units digit. What is the probability that Brian's number is prime? Express your answer as a common fraction.

Solution: Note that all two-digit prime numbers must be odd. Hence, if Brian's number is prime, the four-sided die must show 1 or 3 to guarantee that the number will be odd. We now can write out all possibilities of Brian's odd two-digit number:

$$11, 13, 21, 23, 31, 33, 41, 43, 51, 53, 61, 63$$

Of the twelve odd numbers, we see that 8 of them are prime, namely

$$11, 13, 23, 31, 41, 43, 53, 61$$

To calculate the total number of possibilities without restrictions, we see that there are 6 possible choices for the first digit, and 4 possibilities for the second digit. As a result, there are $6 \times 4 = 24$ two-digit numbers in all. Our probability is then $\dfrac{8}{24} = \boxed{\dfrac{1}{3}}$.

25. Alex has two boxes full of cubes. The first box contains 7 red cubes and 6 blue cubes. The second box contains 8 blue cubes and 5 red cubes. If Alex chooses a cube randomly from each box, what is the probability that both cubes are red? Express your answer as a common fraction.

Solution: There are $7 + 6 = 13$ cubes in the first box, so the probability that a randomly chosen cube is red is $\dfrac{7}{13}$. Similarly, there are $8 + 5 = 13$ cubes in the second box, so there is a $\dfrac{5}{13}$ chance of Alex choosing a red cube from that box. Since we need both cubes to be red, we multiply the probabilities to get $\dfrac{7}{13} \cdot \dfrac{5}{13} = \boxed{\dfrac{35}{169}}$.

26. Trent is at the origin in the coordinate plane. Every minute, he flips a coin. If he gets heads, he walks right one unit. Otherwise, he walks left one unit. After five minutes, what is the probability Trent is at $(1,0)$? Express your answer as a common fraction.

Solution: If Trent is at $(1,0)$ after five minutes, then the number of times he moved right must be one greater than the number of times he moved left. If we let R denote a move to the right and L be a move to the left, Trent's moves must be some arrangement of $LLRRR$. In other words, out of the five moves, we choose 2 of them to be L, and the rest to be R. There are $\dbinom{5}{2} = 10$ ways to do this.

Every minute, Trent has 2 equally weighted choices to move: left or right. Therefore, there are $2^5 = 32$ possibilities total.

Hence, the probability Trent is at $(1, 0)$ after five minutes is
$$\frac{10}{32} = \boxed{\frac{5}{16}}.$$

27. Five people, Allen, Brad, Cameron, David, and Edwin, are standing in a line. David is in the middle, and Brad is not first. Edwin is behind Brad but in front of Cameron. Who is first in line?

Solution: Since Brad is in front of both Edwin and Cameron, Brad must be in the first, second, or third spot. However, we know that Brad cannot be first, and he cannot be third either because David is third. Thus, Brad is second, making Edwin fourth in line and Cameron last. Then the only remaining person, $\boxed{\text{Allen}}$, must be first.

28. Robert, Amber, Charlie, Jasmine, Annie and Patricia are sitting next to each other (in this order) and one of them is holding a movie. The people at the ends are not holding the movie. The people sitting 2 spaces from both ends are not holding the movie. The person holding the movie is no more than 3 spaces away from Patricia. Who is holding the movie?

Solution: We look through the clues one by one to eliminate people in the line. The people at the ends are not holding the movie, so Robert and Patricia are out. Those sitting 2 spaces from both ends do not have the movie, leaving only Amber and Annie. Amber is more than 3 spaces away from Patricia, so $\boxed{\text{Annie}}$ is holding the movie.

Difficult Problems

1. Adam, Bessie, Carlos, Dan, Edwin, Fred, and George are lining up for a photo in two rows: three people in the front, and four people in the back. Bessie and Dan refuse to sit next to each other but Adam and Edwin must be next to each other in the picture. In how many ways can the seven people be arranged?

   Solution: For simplicity, let A denote Adam, B denote Bessie, and so on. We first note that restrictions only apply to A, B, D, and E. Therefore, the other three people can be arranged after the others have determined their positions in $3! = 6$ ways. In addition, for any valid arrangement, we can swap the places of B and D to get another valid arrangement. The same can be applied to A and E.

   So we count the total number of ways to seat A, B, D, and E first (and not caring about the ordering between B and D and A and E). Then multiply our total by $6 \times 2 \times 2 = 24$. To count the seating arrangements for A, B, D, and E, we break this problem down into cases:

   Case 1: B and D are in the back while A and E are in the front.

   Since B and D cannot be next to each other, there are three arrangements for the 4 seats in the back: B - - D , B - D - , and - B - D ('-' denotes an empty seat). Similarly, since A and E must be together, the possible arrangements for the front seats are: A E - and - A E, for a total of 2 arrangements. Thus, this case produces $3 \times 2 = 6$ seatings for A, B, D, and E.

   Case 2: A and E are in the back while B and D are in the front.

   A and E must be together, so there are three ways to seat them (there are three sets of consecutive seats in the row of four in the back). However, B and D have to be separate, so

81

the only possibility is for B and D to be on opposite ends of the front row. Altogether, we have 3 seatings for this case.

Case 3: B, D, A, and E are in the back.

The only possibility is B A E D. Note that we cannot have A E B D or B D A E, as in both cases B and D are together.

Case 4: B and D are in separate rows and A and E are in the front.

Without loss of generality, we can let B be in the back and D be in the front. Since A and E must sit together, D cannot be in the middle front seat. Thus, the front seats must be either D A E or A E D. In the back, B can sit anywhere, so there are 4 possible seatings for the back. Therefore, for this case, we have $2 \times 4 = 8$ valid arrangements.

Case 5: B and D are in separate rows and A and E are in the back.

Without the loss of generality, we can let B in the back and D be in the front. If B is at one of the ends, A and E can sit in any of the two consecutive seats, giving us $2 \times 2 = 4$ possibilities. Otherwise, we must have either - B A E or A E B -. Consequently, there are $4 + 2 = 6$ arrangements for the back row. In the front, there are three choices for D, so we have $6 \times 3 = 18$ possibilities here.

Lastly, we add up all the totals in each case, and multiply by 24 to get our answer:

$$(6 + 3 + 1 + 8 + 18) \times 24 = \boxed{864}$$

2. How many positive integers less than or equal to 2013 have either exactly one or three prime factors in common with 2013?

Solution: We can compute the total number of integers using the Principle of Inclusion-Exclusion (PIE). First, since we want numbers sharing common prime factors, we calculate the prime factorization of 2013 to get $3 \times 11 \times 61$. If 3 is the common prime factor, there are $\frac{2013}{3} = 671$ multiples of 3 (numbers that have a factor of 3). Similarly, there are

$\frac{2013}{11} = 183$ multiples of 11, and $\frac{2013}{61} = 33$ multiples of 61. Adding these up, we get a total of

$$671 + 183 + 33 = 887$$

However, this overcounts several numbers. For example, the numbers that have both a factor of 3 and of 11 are counted twice: once in the multiples of 3 and once in the multiples of 11. The same is true for numbers containing both 11 and 61 or both 3 and 61. Thus, we need to find the number of multiples of 33, 671, and 183 less than or equal to 2013. There are $\frac{2013}{33} = 61$ multiples of 33, $\frac{2013}{671} = 3$ multiples of 671, and $\frac{2013}{183} = 11$ multiples of 183. For each one, we must subtract twice the number of multiples because we do not want to count it at all in our answer (they contain two prime factors):

$$887 - 2(61 + 3 + 11) = 737$$

Unfortunately, we are now undercounting. In particular, we are subtracting the numbers with all three prime factors too many times. There is only one such number, 2013, and it is counted once in each total of the multiples of 3, 11, and 61, but subtracted off twice in each total of the multiples of 33, 671, and 183. This gives us a net count of $3 - 2(3) = -3$. In order for us to count 2013 exactly once, we must add 4 back into the count, to get a final total of $737 + 4 = \boxed{741}$.

3. Chewing gum is prohibited in the Avid classroom, but occasionally some kid chews gum in the room. If the probability that any student is chewing gum is $\frac{1}{3}$, in a room of 6 students, what is the probability more than half of the students are chewing gum? Express your answer as a common fraction.

Solution: We break this problem down into cases based on the number of people chewing gum.

If four people are chewing gum, there are $\binom{6}{4} = 15$ ways to choose the four that chew gum. After selecting the four people, the probability that those four will chew gum and the other two will not is $\left(\frac{1}{3}\right)^4 \left(\frac{2}{3}\right)^2$. Hence, the probability in this case is $\dfrac{15 \cdot 2^2}{3^6}$.

If five people are chewing gum, we have $\binom{6}{5} = 6$ ways to choose the five people. The probability that those five are all chewing gum and the other person is not is $\left(\dfrac{1}{3}\right)^5 \left(\dfrac{2}{3}\right)$, so the probability that there are exactly five chewing gum is $\dfrac{6 \cdot 2}{3^6}$.

The probability that all six people are chewing gum is $\dfrac{1}{3^6}$.

Combining all of our cases, we see that the probability more than half of the students are chewing gum is

$$\frac{15 \cdot 2^2 + 6 \cdot 2 + 1}{3^6} = \frac{60 + 12 + 1}{729} = \boxed{\frac{73}{729}}$$

4. Alice splits the set $\{1, 2, 3, 4, 5, 6, 7\}$ into two disjoint sets such that the the sum of the elements in each of the two new sets are equal. In how many ways can Alice split the sets? (Two sets are disjoint if they do not share a common element.)

Solution: The sum of all of the numbers is $\frac{(1+7)7}{2} = 28$, so each of the two sets must have a sum of 14. We then can list out all of the possibilities:

$$\{7, 6, 1\} \text{ and } \{5, 4, 3, 2\}$$
$$\{7, 5, 2\} \text{ and } \{6, 4, 3, 1\}$$
$$\{7, 4, 3\} \text{ and } \{6, 5, 2, 1\}$$
$$\{7, 4, 2, 1\} \text{ and } \{6, 5, 3\}$$

In total, there are $\boxed{4}$ ways to split the original set.

5. How many ordered triples of positive integers (a, b, c) satisfy $abc = 2014000$?

Solution: We first find the prime factorization of 2014000 as $2^4 \cdot 5^3 \cdot 19 \cdot 53$. Then a, b, and c must all be in the form

$2^w \cdot 5^x \cdot 19^y \cdot 53^z$. In essence, we need to place each of the prime factors into three groups.

To divide up the four factors of 2, imagine four stars and two dividers. We arrange those six items, and each arrangement gives us a unique distribution of the factors. For example, the arrangement $*|| * **$ tells us that a gets 1 factor of 2, b gets zero factors of 2, and c gets 3 factors of 2. There are $\binom{6}{2} = 15$ ways to arrange the six items, so there are 15 ways to split the four factors of 2.

Similarly, to split the 3 factors of 5, consider 3 stars and 2 dividers. There are $\binom{5}{3} = 10$ ways to arrange the objects, so there are 10 ways to distribute the factors.

For both 19 and 53, there are 3 choices to place each factor, either in a, b, or c.

As a result, the total number of ordered triples (a, b, c) is $15 \times 10 \times 3 \times 3 = \boxed{1350}$.

6. Sarah and Anna take turns rolling a 6-sided die, and Sarah rolls first. The person who rolls a 6 first wins and the game ends after that. What is the probability that Anna wins? Express your answer as a common fraction.

Solution 1: We can calculate the probability by considering how many times Anna has to roll to win the game. If Anna only rolls once, she can win if Sarah does not roll a 6 on her first roll and then Anna does. That is, if Anna rolls once, the probability she wins is $\dfrac{5}{6} \cdot \dfrac{1}{6}$.

If Anna has to roll twice to win, the first three rolls (Sarah's first two rolls and Anna's first) must not be 6, and Anna's second roll needs to be a 6. Then the probability is $\dfrac{5}{6} \cdot \dfrac{5}{6} \cdot \dfrac{5}{6} \cdot \dfrac{1}{6}$.

In general, consider Anna's nth roll. There are $2n - 1$ rolls before her winning roll. For Anna to win on the nth roll, all of the previous rolls must not show 6, and the nth roll has to show 6. The probability is $\dfrac{1}{6} \left(\dfrac{5}{6} \right)^{2n-1}$.

We now note that summing the probabilities over all cases

gives us an infinite geometric series:

$$\frac{5}{6} \cdot \frac{1}{6} + \left(\frac{5}{6}\right)^3 \cdot \frac{1}{6} + \cdots + \left(\frac{5}{6}\right)^{2n-1} \cdot \frac{1}{6} + \cdots$$

$$= \frac{\frac{5}{6} \cdot \frac{1}{6}}{1 - \frac{5}{6} \cdot \frac{5}{6}}$$

$$= \boxed{\frac{5}{11}}$$

Solution 2: Let p be the probability Anna eventually wins when it is Sarah's turn. Then after Sarah rolls and does not roll a 6, either Anna rolls a 6 or she doesn't. If Anna does roll a 6, she wins and the game ends. If she doesn't, it is Sarah's turn again, and the probability she wins is again p. Therefore, we can write

$$p = \frac{5}{6} \cdot \frac{1}{6} + \frac{5}{6} \cdot \frac{5}{6} \cdot p$$

We now can solve for p.

$$p = \frac{5}{36} + \frac{25p}{36} \implies \frac{11p}{36} = \frac{5}{36}$$

so $p = \boxed{\frac{5}{11}}$.

7. The expression $99999^3 + 3 \cdot 99999^2 + 3 \cdot 99999 + 1$ can be written equivalently as a^b, where a and b are positive integers greater than 1. What is the minimum value of $a + b$?

Solution: Whenever we are dealing with big numbers, we can replace that big number with a variable. Doing so can give us a better picture of the expression. If we let $x = 99999$, our expression reduces to $x^3 + 3x^2 + 3x + 1$. We then recognize this as the binomial expansion of $(x+1)^3$. Now, we can plug $x = 99999$ back in to get $100000^3 = 1000^5 = 10^{15}$. Clearly, the minimum value of $a + b$ is $10 + 15 = \boxed{25}$.

8. When $(a+2b+3c)^6$ is expanded, it contains the term Nab^2c^3, where N is an integer. Compute N.

Solution: Imagine 6 boxes, each having three items in it: a, $2b$, and $3c$. We want to choose one item out of each box so that we end up with one a, two $2b$'s, and three $3c$'s.

To choose item a, we have 6 choices for which box to choose from. Then for the two $2b$'s, we have five boxes remaining and we pick two of them, giving us $\binom{5}{2} = 10$ choices. Then we must choose $3c$ from each of the last three boxes, which we can do in only one way. Thus, there are $6 \times 10 = 60$ ways to choose the six items. To find the value of N, we multiply all 6 items together with the number of ways to choose them:

$$60 \cdot a \cdot (2b)^2 \cdot (3c)^3 = 60 \cdot 4 \cdot 27 \cdot ab^2c^3 = 6480ab^2c^3$$

Hence, $N = \boxed{6480}$.

9. Five books of heights 5 inches, 7 inches, 8 inches, 9 inches, and 12 inches are to be ordered in a row on a bookshelf. Consecutive books must have a height difference of at most three inches. In how many ways can the books be arranged?

Solution: We first notice that the 12-inch book must be placed at one of the ends. If it were in the middle, the two books next to it would need to have a height within 3 inches from 12. However, only the 9-inch book satisfies this condition. Hence, the 12-inch book must be at one end, with the 9-inch book adjacent to it. There are 2 ways (2 ends of the bookshelf) to do this.

Consider the three remaining books: 5-inch, 7-inch, and 8-inch. Note that the 5-inch book cannot be placed next to the 9-inch book. Thus, there are two ways to pick which of the other two books are next to the 9-inch book. Finally, there are $2! = 2$ ways to arrange the remaining two books (observe that after placing a book next to the 9-inch book, the other books are all within 3 inches of each other). As a result, we have $2 \times 2 \times 2 = \boxed{8}$ ways to arrange the books.

10. Dan is on the first floor of a building. On each floor, he chooses one of the three elevators. One elevator takes Dan up one floor, another takes Dan up two floors, and the last one takes Dan up three floors. In how many ways can Dan get to the eighth floor?

Solution: Notice that once we take an elevator from the first floor, we face the exact same problem: choose one of the three elevators that move up to another floor. We can take advantage of these smaller subproblems by using recursion. Let $f(n)$ is the number of ways Dan can get from the first floor to the nth floor. To get to the nth floor, Dan must take an elevator from either floor $n-1$, $n-2$, or $n-3$. Each of those three floors contains exactly one elevator that gets to floor n. Thus, we see that $f(n) = f(n-1) + f(n-2) + f(n-3)$.

Using this relation, we can determine $f(8)$ by using smaller values of n for $f(n)$. Clearly, $f(0) = 0$ because it is impossible to get to the zeroth floor from the first floor. Also, $f(1) = 1$ since there is one way to get to the first floor from the first floor (do nothing), and $f(2) = 1$ as well. Now,

$$f(3) = f(2) + f(1) + f(0) = 1 + 1 + 0 = 2$$

$$f(4) = f(3) + f(2) + f(1) = 2 + 1 + 1 = 4$$

$$f(5) = f(4) + f(3) + f(2) = 4 + 2 + 1 = 7$$

$$f(6) = f(5) + f(4) + f(3) = 7 + 4 + 2 = 13$$

$$f(7) = f(6) + f(5) + f(4) = 13 + 7 + 4 = 24$$

Finally, $f(8) = f(7) + f(6) + f(5) = 24 + 13 + 7 = \boxed{44}$.

11. Ted's calculator malfunctioned! Now, whenever he presses one of the ten digits (0 - 9), one digit out of the ten is chosen at random and shown on the screen instead. Each digit is equally likely to appear. If Ted presses the buttons 1, 2, and 3 in that order, what is the probability the number that shows up on the screen is greater than 123? Express your answer as a common fraction.

Solution: Ted presses three digits, and all digits are equally likely to come up. This means that any string of three digits $(000, 001, 002, \ldots, 999)$ have the same chance of being shown! Out of those 1000 numbers, there are $999 - 123 = 876$ of them that are greater than 123 (124 through 999). Thus, the probability the number shown is greater than 123 is $\dfrac{876}{1000} = \boxed{\dfrac{219}{250}}$.

12. Eve and Evan choose letters from M, A, T, H, I, S, F, U, N. Each player can choose all, some, or none of the letters, but they cannot choose the same letter. After they choose their letters, they realize that they have chosen all of the vowels. In how many ways could the two have chosen their letters? Note that not all of the letters have to be taken.

Solution: We consider the possibilities of where each letter could end up after Eve and Evan finish choosing. For the consonants, there are three possibilities: picked by Eve, picked by Evan, or not picked at all. We have 6 consonants (M, T, H, S, F, N), so there are $3^6 = 729$ ways the consonants can be picked. However, since all the vowels have to be chosen at the end, there are only two places the vowels can be at the end: with Eve or with Evan. Thus, there are $2^3 = 8$ ways to pick the three vowels. Altogether, there are $729 \times 8 = \boxed{5832}$ ways for the two to pick the letters.

Additional Problems

1. Compute the number of ways to rearrange the word APPLE.

2. How many divisors of 60 are there? A positive integer d is a divisor of 60 if $\frac{60}{d}$ is a positive integer. In particular, 1 and 60 are divisors of 60.

3. Assume every 7-digit whole number is a possible telephone number except those that begin with 0 or 1. What fraction of telephone numbers begin with 9 and end with 0? Express your answer as a common fraction.

4. How many whole numbers between 100 and 400 contain the digit 2?

5. How many ways are there to pick 2 donuts from a box of 12 distinct donuts?

6. Victor is playing a card game, and he is in trouble. In this round, he needs to randomly choose a card from a full deck of 52 cards. He will survive this round if and only if he chooses the queen of spades, or a face card from the suit of hearts. Compute the probability that Victor will not survive this round.

7. If there are 300 cubbyholes and Joe has 301 flyers, what is the probability that one cubby will have at least two flyers?

8. John and Eric are going to a ball game. John is always hungry, so they stop by a vendor that sells sandwiches. The

vendor offers a choice of meat from salami, ham, turkey, or chicken; a choice of cheese from American, mozzarella, and Swiss; and a choice of sauce from barbeque, sour cream, mayonnaise, mustard, and dijon. Each sandwich has a meat, a cheese, and a sauce. Unfortunately, John is a picky eater, so he will not eat sandwiches that have both salami and Swiss, chicken and mustard, or mozzarella and dijon. Compute the number of different sandwich combinations John can order.

9. Ollie the Octopus needs to find socks in the morning. He has bins full of 5 different colors of socks, with each bin having a different color. What is the least number of socks does he have to pull out in the dark so that he can wear eight socks that are all the same color?

10. In Mrs. Smith's class, 19 students play piano, 12 students play violin, and 10 students sing. 5 students play piano and sing, 6 students play both piano and violin, 2 students play violin and sing, and 3 students don't do any music. If her class has 33 students, how many students do all three?

11. 20 couples go to a fancy party. Each person shakes hands with everyone else at the party except for their spouse. Calculate the number of handshakes at the party.

12. A family of 5 members (the mother, father, and three children) are driving to Disneyland in a 5-seat sedan. Only the mother and father can drive, and Bob (one of the three children) becomes carsick when sitting in the passenger's seat. How many comfortable seating arrangements are there?

13. OCMC High School has two classes, Calculus and Shakespearean Literature. 125 students are enrolled in Calculus while 18 students are enrolled in Shakespearean Literature. If 8 students are enrolled in both, and obviously a student

was take at least one of these classes, how many students attend OCMC High School?

14. Kevin fails his biology tests 45% of the time. If Kevin takes 460 tests, on average how many tests will Kevin fail?

15. Compute the number of diagonals in a dodecagon.

16. Farmer Robert has a rectangular orchard that he wishes to fence, of dimensions 400 feet by 500 feet. He plans to buy high-security fence posts, which are sold individually at his favorite store. To ensure that his orchard is securely fenced, he wants to space his fence posts evenly, one at each corner, every 10 feet. Compute the number of fence posts that he must buy.

17. Two dice are thrown. What is the probability that the two-digit number formed will be divisible by 9?

18. Olivia likes to play basketball every other day of the week. If she starts playing on a Tuesday, how many times will she play basketball in two weeks? (The two weeks begin on Tuesday and end on Monday 13 days later.)

19. An 8-sided die is rolled (sides numbered 1-8), and then a 6-sided die is rolled. What is the probability that at least 1 prime number is rolled? Express your answer as a common fraction.

20. Jason and his girlfriend Sally are going on a romantic picnic. Since Jason is very forgetful, he does not know what kinds of sandwiches Sally enjoys. To accommodate for this, he packs 3 types of breads, 5 types of cheeses (including some stinky ones!), and 6 types of meat. However, Sally refuses to have

blue cheese with wheat bread, or white bread with roasted ham. How many choices of sandwiches does Sally have, if she must choose one bread, one cheese, and one meat?

21. Emily and Jane are playing rock-paper-scissors. If wins are worth 1 point, ties are worth half a point, and losses are worth 0 points, in how many ways can Jane get 3 points in the first 4 rounds?

22. Seventeen children are in a class, with 10 boys and 7 girls. If all the girls shake each other's hand once, and all the boys shake each other's hands once, how many handshakes take place?

23. Joey is buying donuts. The store only has 2 boxes each of sprinkled donuts, chocolate donuts, and glazed donuts. He wants to buy 3 boxes. In how many ways can he choose the boxes? (Boxes containing the same type of flavor are indistinguishable and the order in which Joey chooses the boxes does not matter.)

24. Larry the zookeeper needs to feed all the animals. There are monkeys, giraffes, elephants, lions, zebras, and hippos. If he must feed the lions either first or last, in how many different orders can he feed the animals?

25. Compute the number of different ten-digit integers that have 4 zeroes, 2 ones, and 4 twos.

26. Nancy, Sarah, Janice, Frank, Joseph, and Clark are going to attend a lecture by a prestigious professor, and they have saved six consecutive seats in a row. Unfortunately, Nancy and Clark recently had an argument, so they refuse to sit next to each other. Compute the number of possible seating arrangements.

27. A spinner has 24 sectors, numbered $1, 2, 3, \cdots 24$. What is the probability that, after spinning the spinner once, it lands on either a prime number or an odd number? Express your answer as a common fraction.

28. How many possible ways are there to assign true/false values to Statements 1, 2, and 3 such that the system is consistent?

    1. Statement 2 is true
    2. Exactly 2 statements are true
    3. Statement 1 is false

29. William, Kevin, Lohit, and Francis were discussing their possible grades in mathematics class during the grading period. William said, "If I get an A, then Kevin will get an A." Kevin said, "If I get an A, then Lohit will get an A." Lohit said, "If I get an A, then Francis will get an A." All of these statements were true, but only two of the students received an A. Which two received A's?

30. There are 30 people in the OCMC Problem Writing Committee. They wrote some algebra problems, geometry problems, and counting problems. At a meeting, 15 people wrote algebra problems, 15 people wrote geometry problems, and 20 wrote counting problems. One person wrote only counting problems, 2 wrote only geometry, and 3 wrote only algebra. If 8 people wrote both geometry and counting but not algebra, how many people slacked off and didn't write any problems?

31. How many ways are there for 7 people to sit in a row if there are two couples who want to sit with their spouses?

32. Fred the fly is randomly drawing socks out of his drawer without looking. He has 42 red socks, 42 blue socks, 42 green socks, and 13 black socks in his extremely large drawer. How

many socks does he need to take out before he can be sure that he has six socks of the same color?

33. How many ways can you arrange the letters of "Lollipop" if the capital L is different than the 2 lowercase l's?

34. You are playing a game in which you roll a fair six-sided die each round. If you roll a 2, 3, or 5, you get $2, $3, or $5, respectively. If you roll a 4, you lose $4. If you roll a 1 or a 6, you don't lose or gain any money. What is the expected value of playing the game once?

35. How many ways are there to seat 3 girls and 6 boys in a row of 9 chairs if the boys must sit in groups of 3? (There must be girls in between the groups of 3; the boys can't sit in one group of 6.)

36. At a restaurant, for a cheeseburger, you can order either white or wheat bread for the buns; only one type of meat; either cheddar, American, or Swiss cheese; and some, none, or all of 5 toppings: lettuce, tomatoes, onions, pickles, and ketchup. How many different cheeseburgers can you order?

37. Jason is randomly grabbing cookies from a jar that contains chocolate chip cookies, peanut butter cookies, and snicker-doodles. What is the least number of cookies he must grab to ensure he has at least 5 of one kind?

38. There are 3 oranges, 4 apples, 7 bananas, and 1 watermelon in the pantry. How many ways are there to put the fruits in a line, assuming that fruits of the same kind are indistinguishable?

39. When an unfair die with faces 1-6 is rolled, the probability of rolling a number X is $\frac{X}{21}$. When this die is rolled twice, what is the probability that the sum of the rolls is even?

40. 4 people are being split into two groups, each group having two people. In how many ways can they be split?

41. A 3 inch by 3 inch by 3 inch cube is painted on all sides before being cut up into 27 identical cubes. What fraction of the smaller cubes have less than 2 faces painted? Express your answer as a common fraction.

42. Robert flips a coin 4 times. What is the probability that he flips heads at least 3 times? Express your answer as a common fraction.

43. Deborah wakes up in the middle of the night to eat a midnight snack. There's a cookie jar next to her bed. Deborah doesn't know which cookie she'll take out since it's dark, but she does know that  the jar contains 6 chocolate chip cookies, 4 oatmeal cookies, and 5 sugar cookies. What is the minimum number of cookies Deborah needs to remove from the jar to ensure that she has two of a single type of cookie?

44. A fair coin is flipped 5 times. What is the probability that the number of heads that show up will be greater than the number of tails that show up? Express your answer as a common fraction.

Challenge Problems

1. How many ways are there to distribute 11 distinct pieces of candy to 3 children?

2. Eight people, A, B, C, D, E, F, G and H, finish a race. Given that A beat B, B beat C, C beat D, E beat F, F beat G, and G beat H, how many possible overall rankings are there?

3. Jack and Jill go up the hill with 5 empty pails to a place with 3 rivers. Each of the 5 pails can hold 6 gallons of water. What is the shortest time (in minutes) they need to fill all five pails if each of the rivers gives 1 gallon in 1 minute and no river can give water to more than 1 pail at the same time?

4. A group of 10 people are seated in a row of 10 chairs. After leaving their seats for a short break, the people return. In how many ways can the people be seated so that each person is either sitting in her or his previous chair or in a chair adjacent to her or his previous chair?

5. Victor is buying his baguettes. There are 5 different lengths of baguettes, which are 1, 2, 3, 4, and 5 feet. In how many ways can he make a baguette-chain of length 10 feet with these baguettes?

6. A group of 15 balls are in a line. 5 of these balls are red, and the rest are blue. In how many ways can the line be formed such that a red ball is always next to at least one other red ball?

7. Ana and Banana are playing a game. On a player's turn, they spin a spinner with 5 equal sections that are labeled 1, 2, 3, 4, 5. Ana wins if she spins a 3 or higher, and Banana wins if she spins a prime. The game does not end until one player wins. What are the chances that Banana wins if she goes second?

8. A die is rolled until a 5 shows up. What is the expected value of the sum of the rolls?

9. Al and Ben are playing a card game. Each turn, they pick a random card with replacement from the deck of 52 playing cards. The first person to pick an ace wins. If Al goes first, what is the probability that Ben wins?

10. A machine tests a person for poison. The machine has a 5% chance of giving a wrong answer. 1000 people use the machine, 10 of whom are poisoned. If the machine tells a person they are poisoned, what is the probability that they really are?

11. You need to cross a river, but you also need to transport a hungry cat, a hungry mouse, and some cheese. You have one boat, and the boat can only transport you and one other item. If the cat is left with the mouse, the mouse will be devoured. If the mouse is left with the cheese, the cheese will be gone. Compute the minimum number of trips you must make across the river.

In many plants, their petals and leaves follow patterns that relate to the Fibonacci numbers. For example, the lily has 3 petals, buttercups have 5, and daisies have 34. All three numbers are Fibonacci numbers. The spiraling pattern in the seed heads of plants such as sunflowers can also be related to the Fibonacci sequence.

Chapter 4

Number Sense

Warm-up Problems

1. How many numbers in the set $\{1, 4, 9, 7, -5\}$ cannot be the square of a real number?

2. What is the product of the integers in the set

$$\{-15, -22, -12, 16, 0, 17, 195, 57\}?$$

3. Compute the sum of the first 5 prime numbers.

4. What is the greatest integer that divides both 36 and 54?

5. If the natural number x is divisible by 2, 3, 4, 5, and 6, what is the least possible value of x?

6. How many even positive integers divide into 18?

7. Compute the sum of the factors of 60 that are greater than 12.

8. In the month of May there are 31 days. If May 4th is a Sunday, how many Sundays are there in May?

9. What is the sum of all the positive divisors of 24 that are not positive divisors of 18?

Introductory Examples

1. How many perfect squares are there between 2013 and 4026?

   Solution: We can start by finding squares close to 2013 and 4026. We have $45^2 = 2025 > 2013$ and $44^2 = 1936 < 2013$, so the smallest square between 2013 and 4026 is 45^2. Likewise, $64^2 = 4096 > 4026$ and $63^2 = 3969 < 4026$, so 63^2 is the last square in the interval. Between 45 and 63 inclusive, there are $63 - 45 + 1 = 19$ numbers, so there are $\boxed{19}$ such squares.

2. If p and q are primes such that $p + q = 9$, what is the value of pq?

   Solution: If $p + q = 9$, we must have $p, q < 9$ since primes are positive. The only positive primes less than 9 are 2, 3, 5, and 7. Testing each combination, we find that the only solutions are $p = 2$ and $q = 7$ (or vice versa). It follows that $pq = 2 \cdot 7 = \boxed{14}$.

3. How many numbers between 1 and 500 are multiples of 4, 5, and 6?

   Solution: If a number is a multiple of 4, 5, and 6, then it must be divisible by the least common multiple (LCM) of 4, 5, and 6. Since $4 = 2^2$ and $6 = 2 \cdot 3$, we find that the $\text{lcm}(4, 5, 6) = 2^2 \cdot 3 \cdot 5 = 60$. It suffices to find the number of multiples of 60 between 1 and 500. There are $\left\lfloor \dfrac{500}{60} \right\rfloor = \boxed{8}$ such multiples.

4. Farmer Fred is driving down a road and sees cows along the side of the road. Every three minutes after he begins driving, Fred spots a brown cow. Every five minutes after he begins driving, Fred spots a white cow. If Fred drives for twenty-three minutes, how many cows does he spot?

101

Solution: If Fred sees a brown cow every 3 minutes, he will see one at the 3 minute mark, the 6 minute mark, etc. This means that he will spot a brown cow whenever the number of minutes he has driven reaches a positive multiple of three. There are 7 multiples of 3 less than or equal to 23 (the quotient of 23 divided by 3 is 7), so Fred will see 7 brown cows. Similarly, Fred sees a white cow when the number of minutes he has driven reaches a multiple of five. Therefore, he will see 4 white cows because 23 divided by 5 produces a quotient of 4. Altogether, Fred will spot $7 + 4 = \boxed{11}$ cows.

5. If today is Tuesday, what day will it be 251 days from today? (Disregard leap years.)

   Solution: Since we know a week is 7 days, we can try to find how many full weeks there are in 251 days. When 7 is divided into 251, we get 35 and a remainder of 6. This means that after 251 days, 35 whole weeks and 6 days will pass. So after 35 weeks, it will be Tuesday again. Then 6 days later will be $\boxed{\text{Monday}}$.

6. Compute the sum of the last two digits of 25^{2014}.

   Solution: By computing a few powers of 25 one can easily see that the last two digits are always 25, making the sum $\boxed{7}$.

7. Compute: $\dfrac{8! + 6!}{7! + 5!}$.

   Solution: Factoring out 6! from the top and 5! from the bottom, we have $\frac{6!(8 \times 7 + 1)}{5!(7 \times 6 + 1)} = 6 \times \frac{57}{43} = \boxed{\dfrac{342}{43}}$.

8. What is the sum of the distinct prime factors of 2013?

   Solution: 2013 must be divisible by 3, since the sum of the digits of 2013, 6, is divisible by three. So, $2013 = 3 \times \frac{2013}{3} = 3 \times 671 = 3 \times 11 \times 61$. So, the distinct prime factors of 2013 are 3, 11, and 61, so the sum of them is $3 + 11 + 61 = \boxed{75}$.

9. Brian wants to buy a new computer that costs $900. He has saved his birthday money of $150, and he wants to save up the rest of the money by mowing lawns. He can only mow one lawn each day, from Monday to Friday each week. For each lawn he mows, he receives $15. If he starts mowing on Monday, how many days will he have saved enough to buy his computer?

Solution: Brian still needs $900 − $150 = $750 to buy his computer. This means that he will need to mow for at least $750 ÷ $15 = 50 days before getting his computer. Since he starts mowing on Monday, he will need to mow for 10 weeks, so it will take him 70 days, but he will be able to finish getting his money by Friday, so he will only need $\boxed{68}$ days.

10. How many two digit integers have the property that the integer is equal in value to the sum of its digits, multiplied by 4?

Solution: Let us say that this two-digit integer is equal to $10 \cdot A + B$, where A and B are integers, and $A > 0$. Then $10 \cdot A + B = 4(A + B)$. Thus, $6A = 3B$, or $2A = B$, so $A \in \{1, 2, 3, 4\}$. This produces $\boxed{4}$ different two-digit integers satisfying this property.

11. A six digit number is called *awesome* if it is in the form *abcabc*, where *a*, *b*, *c* are digits and $a \neq 0$. For example, 256256 or 678678 are *awesome*. What is the largest integer that divides all *awesome* numbers?

Solution: Observe that $abcabc = abc \cdot 1001$. Thus, 1001 divides into all *awesome* numbers. However, since *abc* can be any three digit number, we cannot guarantee there will be another factor that divides into *abcabc*. For example, the greatest common divisor of 100100 and 101101 is 1001 because 100 and 101 share no common factors other than 1.

103

Hence, the largest integer that divides all *awesome* numbers is $\boxed{1001}$.

12. Perry is on a lifeboat with 400 cans of dried fish, 800 bottles of water, and 200 packets of crackers. How many whole weeks can he survive on this food if he drinks 2 bottles of water per day, eats 2 cans of fish every 3 days, and eats 3 packets of crackers every 7 days?

    Solution: Perry has $\dfrac{800}{2} = 400$ days before he runs out of water and $\dfrac{400}{2} \times 3 = 600$ days before he runs out of fish. He also can live for at least $\left\lfloor \dfrac{200}{3} \right\rfloor \times 7 = 462$ days before he runs out of crackers. Thus, we see that the limiting item is water (only 400 days). This means that Perry will only survive for $\left\lfloor \dfrac{400}{7} \right\rfloor = \boxed{57}$ whole weeks.

13. The perimeter (in inches) of a rectangle with integer side lengths is equal to twice the value of its area (in square inches). What is the perimeter of the rectangle (in inches)?

    Solution: If we let l be the length of the rectangle and w be the width, we have $2(l + w) = 2lw$. Dividing by 2, we have $l + w = lw$. The only positive integer solutions to this equation is $l = w = 2$. To see why this is true, we rewrite the equation as
    $$lw - l - w = 0$$
    $$lw - l - w + 1 = 1$$
    $$(l - 1)(w - 1) = 1$$
    Since l, w are positive integers, we must have $l - 1 = 1$ and $w - 1 = 1$, so $l = w = 2$. As a result, the perimeter of the rectangle is $2(2 + 2) = \boxed{8}$ inches.

14. The product of the first x natural numbers is equal to 192 multiplied by 210. Find x.

Solution: We can try dividing each natural number starting from 1 into 192×210 and see if it divides evenly. If it does, we continue. Otherwise, we have found x. To make computation easier, we prime factorize 192×210 as $2^7 \times 3^2 \times 5 \times 7$. Let $N = 2^7 \times 3^2 \times 5 \times 7$.

Clearly, 1 divides into N. 2 also divides into N, and after we divide N by 2, we are left with $N = 2^6 \times 3^2 \times 5 \times 7$. Continuing, we have

$$\frac{2^6 \times 3^2 \times 5 \times 7}{3} = 2^6 \times 3 \times 5 \times 7$$

$$\frac{2^6 \times 3 \times 5 \times 7}{4} = 2^4 \times 3 \times 5 \times 7$$

$$\frac{2^4 \times 3 \times 5 \times 7}{5} = 2^4 \times 3 \times 7$$

$$\frac{2^4 \times 3 \times 7}{6} = 2^3 \times 7$$

$$\frac{2^3 \times 7}{7} = 2^3$$

$$\frac{2^3}{8} = 1$$

Consequently, we see that $x = \boxed{8}$.

15. If A, B, and C are distinct non-zero digits such that $ABC + 72 = BCA$, compute the sum of all possible values of A.

Solution: Since $A \neq B$, we see that the tens place must carry a 1 to the hundreds place, making $A + 1 = B$. If $C = 7$, then $A = 9$, so $B = 10$, which is not allowed. If $C = 8$, the tens digit place tells us that $1 + B + 7 = 8 + 10$ (carry from units place and carry to hundreds place), so $B = 10$. Similarly, if $C = 9$, we get an invalid solution for B.

However, for $C = 0, 1, 2, 3, \cdots, 6$, we get a unique solution ABC satisfying the criteria. As a result, the sum of all possible values of A is $2 + 3 + \cdots + 8 = \boxed{35}$.

16. For how many positive integers n is $2n^2 + 7n - 15$ prime?

Solution: First, we can factor the quadratic as $(2n - 3)(n + 5)$. A prime number can only have 1 and itself as factors, so either $n + 5$ or $2n - 3$ must be equal to 1. If $n + 5 = 1$, we get $n = -4$, which isn't possible because n must be positive. If $2n - 3 = 1$, we get $n = 2$, so the whole number becomes 7, which is prime. Hence, there is only $\boxed{1}$ positive integer n that makes $2n^2 + 7n - 15$ prime.

17. Determine the units digit of $17^{99} + 13^{99}$.

Solution: We compute powers to find a pattern for the units digit. To find the units digit, we need only consider powers of 7 and 3 due to their being the units digits of their respective numbers. Computing powers of 7, we see that the units digits are $7, 9, 3, 1, 7, 9$, etc. Similarly, the units digits of powers of 3 are $3, 9, 7, 1, 3, 9$, etc. Thus, the units digit of 17 will be 7 if the power is $1 \equiv$ (mod 4), 9 if the power is $2 \equiv$ (mod 4), 3 if the power is $3 \equiv$ (mod 4), and 1 if the power is $0 \equiv$ (mod 4). Similarly, the units digit of 13 will be 3 if the power is $1 \equiv$ (mod 4), 9 if the power is $2 \equiv$ (mod 4), 7 if the power is $3 \equiv$ (mod 4), and 1 if the power is $0 \equiv$ (mod 4). Since $99 \equiv 3$ (mod 4), the units digit of the sum is $3 + 7 = \boxed{0}$ (mod 4).

18. In the world of Aate, people follow the base 8 system. Under the world of Aate lives the Twinkies. The people who live there follow the base 2 system. One day, an Aater goes to the Twinkies seeking trade. He tells the people that he has 14 carrots to trade, but what is the number he should have told the Twinkies so that they would know the real number of carrots in their system?

Solution: Note that the '14' carrots the Aater tells the Twinkies is in base 8. We must convert that to base 2 so the Twinkies can process it. Note that $14_8 = 8 \cdot 1 + 4 = 12$ in base 10. We then convert 12_{10} into base 2, which becomes $\boxed{1100}$ in base 2.

19. Both the base 5 and base 8 representation of a positive integer N are two-digit palindromes. Compute the value of N (in base 10).

Solution: If N is a two-digit palindrome in base 5 and base 8, we can let $N = aa_5 = bb_8$, where a and b are nonzero digits in base 5 and base 8 respectively. We then have

$$5^1 \cdot a + 5^0 \cdot a = 8^1 \cdot b + 8^0 \cdot b$$

which reduces to $6a = 9b$, or $2a = 3b \implies \dfrac{a}{b} = \dfrac{3}{2}$. Clearly, $a = 3$ and $b = 2$ satisfies the proportion. The next smallest solution would be $a = 6$ and $b = 4$. However, since a is a base 5 digit, $a < 5$, so a cannot be 6. Hence, the only solution to the equation is $a = 3$ and $b = 2$. This produces a value of $\boxed{18}$ for N.

20. Convert $.\overline{45}_7$ into a fraction in base 10.

Solution: Given a repeating decimal, $.\overline{3}_{10}$ in base 10, we can express it as an infinite sum of fractions:

$$.\overline{3}_{10} = \frac{3}{10^1} + \frac{3}{10^2} + \frac{3}{10^3} + \cdots$$

Similarly, for other bases, we can 'expand' the repeating decimal. We then have

$$.\overline{45}_7 = \frac{4}{7^1} + \frac{5}{7^2} + \frac{4}{7^3} + \frac{5}{7^4} + \cdots$$

Observe that the fractions with 4 as the numerator form a geometric sequence. Thus, those fractions sum to

$$\frac{4}{7^1} + \frac{4}{7^3} + \frac{4}{7^5} + \cdots = \frac{\frac{4}{7}}{1 - \frac{1}{7^2}} = \frac{28}{48}$$

By the same argument, the fractions with 5 as the numerator also form a geometric sequence, and the sum of all of them is

$$\frac{5}{7^2} + \frac{5}{7^4} + \frac{5}{7^6} + \cdots = \frac{\frac{5}{7^2}}{1 - \frac{1}{7^2}} = \frac{5}{48}$$

Consequently, $\overline{.45}_7 = \dfrac{28}{48} + \dfrac{5}{48} = \dfrac{33}{48} = \boxed{\dfrac{11}{16}}$.

21. For what digit A is $36A958$ divisible by 9?

Solution 1: If a number is divisible by 9, then its residue modulo 9 must be 0. We have

$$36A958 \equiv 3 \cdot 10^5 + 6 \cdot 10^4 + A \cdot 10^3 + 9 \cdot 10^2 + 5 \cdot 10^1 + 8 \pmod 9$$

Since $10 \equiv 1 \pmod 9$ (the remainder when 10 is divided by 9 is 1), we have

$$36A58 \equiv 3 \cdot 1^5 + 6 \cdot 1^4 + A \cdot 1^3 + 9 \cdot 1^2 + 5 \cdot 1^1 + 8 \pmod 9$$

$$\equiv 3 + 6 + A + 9 + 5 + 8 \equiv 9 + A + 9 + 13 \equiv A + 4 \pmod 9$$

As a result, for $36A958$ to be divisible by 9, $A + 4$ must be divisible by 9. Since A is between 0 and 9 inclusive, the only possible value for A is $\boxed{5}$.

Solution 2: A number is divisible by 9 if and only if the sum of its digits is divisible by 9. The sum of the digits of $36A958$ is $3 + 6 + A + 9 + 5 + 8 = A + 31$. Since A is a digit, the only value for A that makes $A + 31$ divisible by 9 is $\boxed{5}$.

22. Adam goes to a market to buy apples and bananas. One apple costs 3 dollars while one
banana costs 5 dollars. If Adam has 72 dollars to spend and must buy at least one apple
and at least one banana, what is the maximum number of apples Adam can buy? Note: Adam must
spend all his money.

Solution: Let a be the number of apples Adam buys, and b be the number of bananas he buys. We have $3a + 5b = 72$, and a and b are positive integers. Taking the equation modulo 5, we see that $3a \equiv 2 \pmod 5$. Multiplying the congruence by 2 and simplifying yields $a \equiv 4 \pmod 5$. This means that as

108

long as a leaves a remainder of 4 when divided by 5, we get a valid solution for a and b.

Because we know that $b \geq 1$, we have $5b \geq 5$, so $3a = 72 - 5b \leq 72 - 5 = 67$. It follows that $a \leq 22$. The largest value of a such that $a \leq 22$ and $a \equiv 4 \pmod 5$ is $a = \boxed{19}$.

Difficult Problems

1. What is the product of the all the factors of 400?

   Solution: Consider a divisor, d, of 400. Then $\dfrac{400}{d}$ is also a divisor of d, and $d \cdot \dfrac{400}{d} = 400$. This means that we can pair off factors of 400 so that the numbers in each pair multiply to 400. Since $400 = 2^4 \cdot 5^2$, there are $5 \cdot 3 = 15$ factors in all. Since the factor 20 pairs with itself, we exclude that from our count for now. Therefore, we have 7 pairs, giving us a product of $400^7 = 20^{14}$. Multiplying this by the factor 20, we get $\boxed{20^{15}}$. (Notice that the answer is equivalent to $400^{\frac{15}{2}}$).

2. Let p be a prime number. If exactly two of the numbers in the set
$$\{p, p+2, p+4, p+8, p+14, p+50\}$$
are prime, compute the second smallest value of p.

   Solution: We can go through the primes one by one..

   If $p = 2$, we get $\{2, 4, 6, 10, 16, 52\}$. Only 2 is a prime number.

   If $p = 3$, we get $\{3, 5, 7, 11, 17, 53\}$. All six of them are primes.

   If $p = 5$, we get $\{5, 7, 9, 13, 19, 55\}$. There are four primes: 5, 7, 13, and 19.

   If $p = 7$, we get $\{7, 9, 11, 15, 21, 57\}$. This prime works, as there are only two primes: 7 and 11.

   If $p = 11$, we get $\{11, 13, 15, 19, 25, 61\}$, producing 4 primes: 11, 13, 19, and 61.

   If $p = 13$, we get $\{13, 15, 17, 21, 27, 63\}$. There are two primes: 13 and 17.

   Thus, the second smallest value of p is $\boxed{13}$.

3. Let $a_1, a_2, ..., a_{2012}$ be a sequence where $a_1 = 2^2 \cdot 3^{10} \cdot 5^{60}$ and a_{i+1} equals the number of factors of a_i for $i \geq 1$. Compute the value of

$$\frac{1 + a_2 + a_4 + a_6 + ... + a_{2012}}{2}.$$

Solution: Let $f(n)$ denote the number of factors of n. We can first compute a few of the first a_i, $i > 1$:

$$a_2 = f\left(2^2 \cdot 3^{10} \cdot 5^{60}\right) = 3 \cdot 11 \cdot 63$$

$$a_3 = f(3 \cdot 11 \cdot 63) = 2 \cdot 2 \cdot 2 = 2^3$$

$$a_4 = f(2^3) = 4 = 2^2$$

$$a_5 = f(2^2) = 3$$

$$a_6 = f(3) = 2$$

$$a_7 = f(2) = 2$$

Note that $a_i = 2$ for $i > 5$ because the number of factors of 2 is 2. Hence,

$$\frac{1 + a_2 + a_4 + a_6 + \cdots + a_{2012}}{2}$$

$$= \frac{1 + 3 \cdot 11 \cdot 63 + 4 + 2 + 2 + \cdots + 2}{2}$$

$$= \frac{1 + 2013 + 4 + 2(1004)}{2} = \frac{4026}{2} = \boxed{2013}$$

4. Compute the smallest integer x such that $(1)(1!) + (2)(2!) + \cdots + (14)(14!) < x!$.

Solution: Rather than computing the whole expression on the left side of the inequality, we can look at simpler cases first:

$$(1)(1!) = 1 < 2!$$

$$(1)(1!) + (2)(2!) = 5 < 3!$$

111

$$(1)(1!) + (2)(2!) + (3)(3!) = 23 < 4!$$

Studying the examples above, it seems that $(1)(1!)+(2)(2!)+\cdots+(n)(n!) = (n+1)! - 1$. To see why this is true, consider the equivalent expression of $(1)(1!) + (2)(2!) + \cdots + (n)(n!)$:

$$(2-1)(1!) + (3-1)(2!) + \cdots + (n+1-1)(n!)$$

$$= (2! - 1!) + (3! - 2!) + \cdots + ((n+1)! - n!)$$

$$= (n+1)! - 1!$$

Now, we can clearly see that the smallest value of x is $14+1 = \boxed{15}$.

5. The number 16 has 2 even perfect square factors. How many even perfect square factors does 720 have?

Solution: We first find the prime factorization of 720, as we often do in problems dealing with factors. We have $720 = 2^4 \cdot 3^2 \cdot 5$. Consider an even perfect square factor of 720. Note that it must be in the form $2^a \cdot 3^b \cdot 5^c$, where a, b, and c are even and $a > 0$ (factor must be even). For a, we have 2 choices: $a = 2$ or $a = 4$. For b, we have two choices: $b = 0$ or $b = 2$. We only have one choice for c: $c = 0$. Therefore, there are $2 \times 2 \times 1 = \boxed{4}$ even perfect square factors of 720.

6. 3 more than the product of positive integers x and y is equal to the sum of $4x$ and y increased by 18. If $x > y$, what is the product xy?

Solution: Rewriting the problem into an equation, we have $3 + xy = 4x + y + 18$. Rearranging, we get $xy - 4x - y = 15$. Adding 4 to both sides, we have $xy - 4x - y + 4 = 15 + 4 = 19$. Observe that the left side can be factored as $(x-1)(y-4)$ (this method of factoring is known as Simon's Favorite Factoring Trick, or SFFT).

Since x and y are positive integers and $x > y$, we must have $x - 1 = 19$ and $y - 4 = 1$. This leads to $x = 20$ and $y = 5$, so $xy = \boxed{100}$.

7. Michael has N marbles, and he wants to arrange them in groups. When he arranges his marbles in groups of 10, then there are 5 left over. When he arranges his marbles in groups of 4, there is 1 left over. If he has more than 60 marbles, what is the least amount of marbles he can have?

Solution: Having 5 marbles left over when arranging the marbles in groups of 10 means that $N \equiv 5 \pmod{10}$. Similarly, if Michael has one marble remaining after grouping them into fours, $N \equiv 1 \pmod 4$. We have two linear congruences, so we can combine them.

The first congruence is equivalent to $N = 10a + 5$, and the second one is the same as $N = 4b + 1$. Therefore, $10a + 5 = 4b + 1$. Taking this equation modulo 4, we get $2a + 1 \equiv 1 \pmod 4$, so $2a \equiv 0 \pmod 4$, or $a \equiv 0 \pmod 2$. As a result, we can let $a = 2c$, so $N = 10(2c) + 5 = 20c + 5$.

Since Michael has over 60 marbles, $20c + 5 = N > 60$. The smallest integer c would be $c = 3$ so $N = 20(3) + 5 = \boxed{65}$. .

8. The number $N = 12345...20112012$, is formed by writing the first 2012 natural numbers together. What is the remainder when N is divided by 9?

Solution: The remainder when a number k is divided by 9 is the same as the remainder when the sum of the digits of k is divided by 9. Thus, N modulo 9 is the same as

$$1 + 2 + 3 + \cdots + 2 + 0 + 1 + 1 + 2 + 0 + 1 + 2 \quad \pmod 9$$

However, we don't want to deal with the sum of every digit of the numbers from 1 to 2012. Instead, we can 'expand' N like this

$$N \equiv 1 + 2 + 3 + \cdots + 2010 + 2011 + 2012 \quad \pmod 9$$

This makes it easier to calculate the answer, as we can add the numbers from 1 to 2012 quickly.

$$1 + 2 + 3 + \cdots + 2012 = \frac{2013 \cdot 2012}{2} = 2013 \cdot 1006$$

As a result,

$$N \equiv 2013 \cdot 1006 \equiv (2 + 0 + 1 + 3)(1 + 0 + 0 + 6) \pmod{9}$$

$$N \equiv 6 \cdot 7 \equiv \boxed{6} \pmod{9}$$

9. Determine the sum of the two smallest prime divisors of $3^8 - 1$.

Solution: Notice that $3^8 - 1$ is the difference of two squares, so we can factor it as $(3^4 - 1)(3^4 + 1)$. Again, $3^4 - 1$ is the difference of two squares. Factoring again, we get $3^8 - 1 = (3^2 - 1)(3^2 + 1)(3^4 + 1) = (8)(10)(82)$. Clearly, both 2 and 5 divide into $(8)(10)(82)$. However, 3 does not because $3^8 - 1 \equiv -1 \pmod 3$. As a result, the two smallest prime divisors are 2 and 5, totaling to $\boxed{7}$.

10. Find the units digit of $2014^{2014^{2014}}$.

Solution: To find the units digit, we consider the number modulo 10. Note that $2014 \equiv 4 \pmod{10}$, so we must find the units digit of $4^{2014^{2014}}$.

Consider the units digit of powers of 4s. We have $4^1 \equiv 4 \pmod{10}$, $4^2 \equiv 6 \pmod{10}$, $4^3 \equiv 4 \pmod{10}$, $4^4 \equiv 6 \pmod{10}$, and so on. Thus, when the exponent is odd, the units digit is 4. Otherwise, the units digit is 6. Obviously, 2014^{2014} is even, so the units digit of the number must be $\boxed{6}$.

11. What is the smallest positive integer that has twice the number of factors that 12 has?

Solution: Let N be the number we are looking for. We first find the number of factors 12 has. Since $12 = 2^2 \cdot 3$, we see that all factors of 12 are in the form $2^a \cdot 3^b$, with $0 \leq a \leq 2$

and $0 \le b \le 1$. Thus, 12 has $3 \times 2 = 6$ factors, so N must have 12 factors.

Considering the prime factorization of N, we see that N must be in the form

$$p_1^{11} \ , \ p_1^5 \cdot p_2 \ , \ p_1^3 \cdot p_2^2 \ , \ p_1^2 \cdot p_2 \cdot p_3$$

where p_1, p_2, p_3 are all distinct primes. If $N = p_1^{11}$, the smallest N would be $2^{11} = 2048$. If $N = p_1^5 \cdot p_2$, the smallest N would be $2^5 \cdot 3 = 96$. If $N = p_1^3 \cdot p_2^2$, N would be at least $2^3 \cdot 3^2 = 72$. Finally, if $N = p_1^2 \cdot p_2 \cdot p_3$, the smallest N can be is $2^2 \cdot 3 \cdot 5 = 60$. Hence, the smallest positive integer with 12 factors is $\boxed{60}$.

12. What is the sum of all even factors of 120?

Solution: We first prime factorize 120 as $2^3 \cdot 3 \cdot 5$. We see that every even factor of 120 is in the form $2^a \cdot 3^b \cdot 5^c$, where $1 \le a \le 3$, $0 \le b \le 1$, and $0 \le c \le 1$. Therefore, note that when we construct some even factor, we first choose the number of factors of 2 it will contain, then how many factors of 3 it will contain, and lastly how many factors of 5 it will contain. The sum of the even factors can then be expressed as
$$(2^1 + 2^2 + 2^3)(3^0 + 3^1)(5^0 + 5^1)$$

Observe that every possible even factor is contained in the expansion of the expression above. Thus, the sum of the all even factors of 120 is $(2 + 4 + 8)(1 + 3)(1 + 5) = (14)(4)(6) = \boxed{336}$.

13. Compute the number of integers x from 1 to 20 inclusive such that $x^2 - x + 1$ is divisible by 7.

Solution: We are trying to solve $x^2 - x + 1 \equiv 0 \pmod 7$. Notice that $x^2 - x + 1 \equiv x^2 - x - 6 \equiv (x - 3)(x + 2) \pmod 7$, so $x \equiv 3 \pmod 7$ or $x \equiv -2 \equiv 5 \pmod 7$. Hence, a complete list of the solutions to this equation can easily be found: $3, 10, 17, 5, 12, 19$ so the number of solutions is $\boxed{6}$.

Additional Problems

1. A monkey wants to climb a very tall tree of height 400 feet in the forest to get to the bananas at the top. Every day, it climbs 30 feet, but it slides down 10 feet during the night when it is resting. How many days will the monkey take to secure a banana?

2. The digits $2, 3, 4, 7,$ and 9 are used to form the smallest possible five-digit even integer. Compute the value of the integer modulo 100.

3. How many even three-digit numbers are divisible by 13?

4. If x is the greatest prime factor of 36 and y is the greatest prime factor of 27, what is $x + y$?

5. Annie is playing games of solitaire and listening to music at the same time. Each game of solitaire lasts for 5 minutes. She only wins a game if she hears a Justin Bieber song for some portion of that game. If the Justin Bieber song will play 30 minutes after she begins playing solitaire, Annie will win for the first time when she plays her nth game. What is the value of n?

6. Bob has a pile of (at least two) candies that he wants to separate into smaller piles of equal size. When he separates them into piles of 3, he has one candy left over. When he puts them into piles of 5, he has one candy left over. When he puts them into piles of 4, he has one candy left over. What is the minimum number of candies that could have been in the original pile?

7. The Fibonacci sequence begins with the terms

$$1, 1, 2, 3, 5, 8, 13, 21, 34, \cdots$$

and each term after the 2nd is defined to be the sum of the previous two terms. What is the remainder when the 1000th term is divided by 5?

8. The sum of 3 consecutive nonnegative integers is equal to the sum of 4 consecutive nonnegative integers. Compute the smallest possible value of this sum.

9. For integers a and b, let $a \heartsuit b = GCD(a, b) \times LCM(a, b)$, where GCD is the greatest common divisor function and LCM is the least common multiple function. Compute: $6 \heartsuit (14 \heartsuit 32)$.

10. If $29x + 11y = 1$ where x and y are integers and $x > 0$, compute the largest possible value of y.

11. A terrible analog clock loses 5 minutes per hour. If the correct time right now is coincidentally shown on the clock, after how many hours will the clock display the correct time again?

12. In a certain year, January had exactly four Tuesdays and four Saturdays. On what day of the week did January 1 fall that year?

13. For every dress that a seamstress sews, she earns \$8. She is given an extra \$4 for every 7 dresses she sews. How many dresses must she sew to earn \$520?

14. Some students are seated at a table when a bag full of $10,000$ candies is passed. If Bob gets both the first piece of candy

and the last piece of candy, what is the maximum possible number of students seated around the table?

15. What is the sum of the digits of the sum of the digits of the sum of the digits of 2^{18}?

16. Compute the number of zeroes at the end of $1^1 \cdot 2^2 \cdot 3^3 \cdot 4^4 \cdots 42^{42} \cdot 43^{43}$.

17. What is the smallest positive integer with four factors?

18. In a total of 15 numbers, the median is 38. The largest number is 93. If the largest number is changed to 97, what is the median?

19. What is the greatest positive integer that divides both $37^2 + 1$ and $37^3 + 37^2 + 1$?

20. How many integers are factors of 1008, but not of 36?

21. What is the smallest integer greater than 3 that leaves a remainder of 3 when divided by 4, 5, 6, and 7?

22. If the digits of a certain two digit integer are reversed, the new number is 36 more than the original integer. The sum of the digits of the original number is 10. What is the new number?

23. What is the value of 6 times 9 in base 10 when expressed in base 13?

24. Harry the snail is climbing over a small hill. Every day, he moves forward 6 inches, and every night, he slides down 2 inches in his sleep. How many days would it take him to climb up one side and down the other side if the hill is 10 feet tall?

25. What is the remainder when the 55th Fibonacci number is divided by 4, assuming that the first Fibonacci number is 1?

26. Victor is buying baguettes. He wants to buy a number of baguettes that would have 2 left over if split among 3, 1 left over if split among 5, 3 left over if split among 8, and 9 left over if split among 11. What is the smallest number of baguettes he can buy?

27. When a certain number not divisible by 10 between 5000 and 6000 exclusive is multiplied by 9, its last three digits remain the same. What is the number?

28. Is 8573624457820152763495 divisible by 9? If not, what is the remainder?

29. If a, b, and c are positive integers and $abc = 12$, how many possible values of ab are there?

30. The mean, median, and mode of the following set are all identical: $17, n, 13, 17, 13, 17, 22$. What is n?

31. Elizabeth has an older brother and older sister. The sum of the three siblings' ages is 8. The product of their ages is 12. How old is Elizabeth?

32. The median of a set of five positive integers is 7. The mean of the set is also 7. The mode of the set is 3. What is the greatest number that could be in the set?

33. If Jerry takes a one-day break from school after every 2 school days, Robi takes a one-day break after every 3 school days, and school starts on a Monday, what is the first day both Jerry and Robi will be both take a one-day break on the same day? Assume that there are no weekends, holidays, etc.

Challenge Problems

1. Let N be the product of all positive integers that divide into 600 but not 420. How many terminal zeroes does N have?

2. Let n be a positive integer less than 40 that has more than 4 factors but less than 9 factors. Compute the number of possible values that n can take on.

3. Compute the first prime k such that 1 more than the product of all the primes less than or equal to k is not prime.

4. When the binomial coefficient $\binom{125}{32}$ is written out in base 10, how many zeroes are at the rightmost end?

5. What is the units digit of $2013^{2014^{2015}}$?

6. In the world of Orange, peels are worth 4 cents and tangerines are worth 7 cents. What is the minimum total number of peels and tangerines needed to make 50 cents?

7. For how many positive integers n ≥ 2 is $n^2 - 3n + 2$ a prime number?

8. The sum of the reciprocals of three consecutive integers is equal to $\frac{47}{60}$. What is the smallest of these three integers?

9. How many of the factors of 237,600 are perfect squares?

10. A positive integer has 20 factors. What is the sum of the smallest 2 numbers that fit that condition?

11. How many factors of 19404 are squares?

12. Eric is driving around his neighborhood. He looks at his odometer, which displays a 5-digit palindrome less than 20,000. 11 miles later, it displays another palindrome. What is his odometer reading 15 miles after that?

The leaves of a ferns follow a pattern that resembles a fractal. Each large stem splits into small stems, which split of into more stems with leaves, which in turn split into large and small veins. Using these ideas, scientists and artists can apply fractals to create accurate computer representations of trees, plants, and other natural objects.

Chapter 5

Review

Problems

1. Evaluate: $4^2 - 2 \times 4 \times 3 + 3^2$.

2. A circular disk with radius 1 is placed randomly on a circular table with radius 7 so that the center of the disk is on the table. What is the probability that the whole disk is on the table, with no part hanging off?

3. Victor is eating his baguettes. He takes 3 inch bites or 5 inch bites. How many ways can he get through a 2 foot baguette?

4. If $a = 32$, $b = 21$, and $c = 29$, what is $(a + c)(a - b)(a - c)$?

5. If the sum $1 + 2 + \cdots + n = 1540$, what is the value of n?

6. How many integers from 1 to 500, including 1 and 500, are multiples of 2 but not multiples of 5?

7. A coin has a probability of $\frac{1}{3}$ that it will show heads when flipped, and a probability of $\frac{2}{3}$ that it shows tails. The coin is flipped twice. What is the probability that it showed tails both times? Express your answer as a common fraction.

8. Jane and Suzy are playing basketball. Jane makes her free throws $\frac{2}{3}$ of the time while Suzy only makes hers $\frac{1}{2}$ of the time. If Jane shoots 2 free throws and Suzy shoots 3, what is the probability that they don't make any free throws? Express your answer as a common fraction.

9. Triangle ABC has a right angle at B. D is the foot of the altitude from B. If BD is 12 and $\frac{AB}{BC} = \frac{3}{4}$, what is AB?

10. Suppose Jennifer writes a list of 100 positive integers less than 100. If she writes another list with each number from

the first list subtracted from 100, then what is the sum of the numbers on both of her lists?

11. In Mr. Faraday's classroom, everyone has at least one calculator. The only types of calculators are TI-84's and TI-89's. Eighteen students have TI-84's, and two-thirds of all the students have TI-89's. If there are 30 students in Mr. Faraday's class, how many students have both a TI-84 and a TI-89?

12. A number is called *sphinxlike* if it has 3 digits and the first and last digits are identical. How many *sphinxlike* numbers are there?

13. Ann and Brenda are having a picnic. They want to make enough sandwiches to feed the both of them without having leftovers. Ann needs to eat 3 sandwiches but can stuff herself to eat 5. Brenda needs to eat 4, but can stuff herself to eat 6. What is the difference between the largest and smallest amount of sandwiches they can bring?

14. John's computer has 2 GB of RAM (1 GB = 1000 MB). His computer requires 509 MB for operating system needs, and he is currently running his internet browser, which takes up 129 MB. He is also running a game that takes up 856 MB. How much RAM is still open?

15. Tom's batting averages for his first three seasons are 0.267, 0.243, and 0.256. The mean of his batting averages for his first four seasons is 0.261. What is his batting average for his fourth season? Express your answer as a decimal rounded to the nearest thousandth.

16. Aaron teaches tennis lessons and earns 200 dollars a week. However, he has to restring his racquets at the end of every two weeks and this costs him 40 dollars. He wants to buy a 920 dollar laptop. If he starts out with 0 dollars, how many weeks will this take him?

17. A sports store sells basketballs and baseballs. Each basketball costs 10 dollars and each baseball costs 6 dollars. Avery buys 13 basketballs and baseballs altogether at the store, and paid 110 dollars total. How many basketballs did he buy?

18. Every student in the All-Girls Tournament shakes hands with every other student, for a total of 153 handshakes. How many students are in the tournament?

19. Two dice are thrown. What is the probability that the sum of the numbers on the top faces is even? Express your answer as a common fraction.

20. In the month of April there are 30 days. How many possible ratios are there of the number of Mondays to the number of Thursdays?

21. $\triangle ABC$ is a right triangle with integer side lengths and hypotenuse BC. Point D is chosen such that A and D are on opposite sides of BC and $BD = CD = 8$. If $BC = 10$, find the perimeter of $ABDC$.

22. If Sally can wash a car in 1 hour, and Victor and Sally together can wash a car in 20 minutes, how long does it take Victor to wash a car by himself?

23. There are 498 students in your grade. It is time to take a class field trip to the Smithsonian Museums, and your teachers attempt to divide the class into equal groups such that each group has only one teacher. However, there are some students that will not be able to be placed into groups. If there are 24 teachers, at most how many students can be in a group?

24. A wood block has a density of 1.5 grams per cubic centimeter. Given that density is mass divided by volume, what is the mass, in grams, of 200 cubic centimeters of this material?

25. Andy picked 6 apples from his apple tree. He cut each apple into 8 slices. When Andy wasn't looking, Alexis came and ate 15 apple slices. If Andy had been planning to give out the apple slices equally among 3 people, how many fewer slices does each person get?

26. A car traveling 95 feet per second is traveling how many miles per hour? There are 5280 feet in 1 mile. Express your answer to the nearest whole number.

27. Daisy and Jane are racing. If Jane gets a head start of 5 seconds, and Daisy catches up to Jane after running 60 feet in 10 seconds, then what was Jane's average speed in feet per second?

28. A list of 300 numbers starts with 1. After that, every number is three more than the number before it. What is the 200th number on the list?

29. If a is 50% of b, and b is 30% of c, what is the value of $\frac{a}{c}$? Express your answer as a common fraction.

30. Allen and Bob have pieces of candy. If Allen gave Bob 3 of his candies, the two boys would have the same amount of candy. Given that the boys have 40 pieces of candy in total, how much candy does Bob have?

31. The equation used to convert C degrees Celsius to F degrees Fahrenheit is $F = \frac{9}{5}C + 32$. How many degrees Celsius is 113 degrees Fahrenheit?

32. Alice went into a fruit store and bought 21 pieces of fruit. The fruit store sells only apples and pears. Apples cost 12 cents each while pears cost 8 cents each. If her total cost was \$2.12, how many apples did Alice buy?

33. What is the value of $99 - 98 + 97 - 96 + 95 - 94 + \cdots + 3 - 2 + 1$?

34. How many ways are there to arrange 5 different keys on a keychain? Two arrangements are considered the same if one can reach the other through rotation and/or reflections.

35. The 600 students at a middle school are divided into three groups of equal size for lunch. Each group has lunch at a different time. A computer randomly assigns each student to one of three lunch groups. What is the probability that three friends, Kevin, Francis, and William, will be assigned to the same lunch group? Express your answer as a common fraction.

36. Compute: $\dfrac{13 + 26 + 39 + \cdots + 1300}{17 + 34 + 51 + \cdots + 1700}$.

37. Eli spends 20% of his salary on food and spends 32% of his salary on housing. If he spends \$585 on food and housing, what is Eli's salary (in dollars)?

38. Billy Bob has 10 songs on his jPod Nano, 90 songs on his jPhone, and 30 songs on his jPod Touch. The 10 songs in his jPod Nano are also included in the 30 songs in his jPod Touch and the 90 songs on his jPhone. How many distinct songs does Billy Bob have in all of the devices?

39. There are 20 chapters in the math book *Math*, and the nth chapter corresponds with the index number of the nth odd number. For example, the first chapter's index number is 1, second chapter is 3, third chapter is 5, and so on. There are as many subsections in each chapter as the corresponding index number. How many subsections are there in the book?

40. There are 900 students in school. 720 take Spanish. Three-fourths of the remaining students take French and the rest take Latin. How many students take Latin?

41. Erik can finish a 20-question quiz in 10 minutes. Kathy can finish the same quiz in 15 minutes. If the two of them work together on the same 20-question quiz, in minutes, how long will it take them to finish?

42. If $x + y + z = 8$, where $x, y,$ and z are nonnegative integers, compute the number of ordered triples (x, y, z).

43. What is the ratio of the number of 9 digit palindromes to the number of 8 digit palindromes? Recall that palindromes are numbers that read the same forward as backwards, and cannot have leading zeroes.

44. A particular brand of lemonade is composed of 75% lemon juice and 25% water. I take 8 oz. of this lemonade and pour it into a beaker with 2 oz. of fresh water. What percentage of the fluid in the beaker is lemon juice?

45. Sarah is delivering newspapers on a street of houses numbered 1 to 100 with even-numbered houses on the right and odd-numbered houses on the left. If she only delivers to house numbers that are a multiple of 3 and on the right, how many houses does she deliver newspapers to?

46. Paul is not very good at talking to girls. Every time Paul tries to talk to a girl, he gets very nervous and starts counting in multiples of 3. If Paul counts in multiples of 3 (3, 6, 9, 12, 15,...) until he reaches 1,242, how many numbers has Paul said?

47. Ryan ran 100 meters from the start line to the finish line in 15 seconds. He then ran back (from finish line to start line) at a pace of 5 meters per second. What was his average speed in meters per second of his 200 meter run? Express your answer as a decimal rounded to the nearest tenth.

48. Alice, Allison and Ali are all siblings. Alice is twice as old as Ali, and the sum of the ages of Alice and Allison is 20. If Ali is 8, how old is Allison?

49. Five mean monkeys are jumping on a bed. After every 10 minutes, the monkeys on the bed randomly and unanimously decide to kill one of their group. What is the probability that monkey Bob survives? Express your answer as a common fraction.

50. A point (x, y) is a *lattice* point if both x and y are integers. Given a *lattice* point (x, y), its *adjacent* lattice points are $(x+1, y)$, $(x-1, y)$, $(x, y+1)$, and $(x, y-1)$. A grasshopper starts at $(0,0)$ on a coordinate plane. Each second, he jumps to an adjacent lattice point. How many possible points can the grasshopper move through or reach after 6 seconds?

51. Simplify the expression $\frac{2^2 \cdot \sqrt{4^3}}{12^2}$. Express your answer as a common fraction.

52. If an ant crawls around the outside of a square with side length 1 inch, always keeping 1 inch away from the boundary of the square, then how many inches will the ant travel in one rotation around the square? Express your answer in terms of π.

53. A turkey escapes from Holly's farm and starts running away in a straight line at 4 meters per second. 10 seconds later, Holly notices the turkey is missing and chases after it at 8 meters per second. How far will Holly run before she catches up to the turkey?

54. Peter can solve *The Last Olympiad* problems in 3 hours, and Annabelle can solve all of them in 4 hours. How many hours will it take Peter and Annabelle to finish all the problems in *The Last Olympiad*? Express your answer as a common fraction.

55. Define the function $a\#b$ as $2a + 2b$. If $x + y + z = 3$, what is the value of $\left(\frac{1}{2}(x\#y)\right)\#z$?

56. An astronaut goes to an alien planet where half of the citizens always tell the truth, and the others always lie. The astronaut meets three citizens, and asks the first one if the second one is a liar. It says, "No, it is not". Then the astronaut turns and ask the second one if the third one is a truth-teller. It says "No, it is not". Finally, the astronaut asks the third one if it is a liar. The third one says, "No, I am not". A truth-teller walks by and says, "1 lie has been told." What are each of the three citizens?

57. Todd lives in a magical pond, where all the toads and frogs talk. One species always says the truth, while the other species always lies. Figure out the species of each talker.

    Todd: My neighbor Tammy is a frog.

    Tammy: My neighbors Todd and Ted are both toads.

    Ted: I am a frog or a toad.

    Terry: Todd is a toad.

58. Maddy wants to buy school supplies. She visits a store that sells pencils in packs of seven, pens in packs of six, and erasers in packs of four. She wants to buy the same number of pencils, pens, and erasers, but can only purchase packs of them. What is the least possible number of pencils Maddy needs to buy?

59. How many two-digit numbers are divisible by 11 and not even?

60. John's mom is 30 years older than John. In 9 years, John's mom will be three times older than John. How old is John's mom now?

61. The distance between town A and town B is 150 miles. Andy is driving from A to B in a car that goes 60 miles per hour. However, there is a 30 mile stretch of road that is jammed with traffic. Traffic decreases the speed of the car by 80 percent. How long will it take Andy to get there, and what will be his average speed?

62. Mister, Miss, and Junior Rabbit are eating out of a rabbit garden with 240 carrots in it. Mister Rabbit eats 24 carrots per hour. Miss Rabbit eats 36 carrots per hour. Junior Rabbit eats 20 carrots per hour. All rabbits eat at only half their speed after noon. If the three rabbits start eating out of the rabbit garden at $10:00$ AM, by what time will they have finished all the carrots in the garden?

63. Daniel has 20 coins, all of which are either pennies or nickels. If the total amount of Daniel's coins is $0.32, how many nickels does Daniel have?

64. In a xy-plane, let A be the point $(1, 1)$, B be the point $(3, 5)$, and C be the point $(4, 1)$. Point B is reflected across the x-axis to produce Point D. What is the area of $\triangle ACD$ in square units?

65. City A is 11 miles from City B. City C is directly south of City B and east of City A. If the distance from City A to City C is 7 miles, what is the distance from City B to City C? Express your answer in simplest radical form.

66. Eugene has a jar of coins that consists of only nickels, dimes, and quarters. He has 24 nickels and dimes, 30 dimes and quarters, and 32 nickels and quarters. How many coins does he have in his jar?

67. Compute the number of ways to order the integers from 1 to 12 such that the even numbers must be in ascending order and the odd numbers must be in descending order.

68. Three animals are arguing about a missing cookie. Frog says that Rabbit ate it, Rabbit says that Frog ate it, and Bird says that Frog is innocent. If exactly 2 animals are lying, who is sure to be innocent?

69. Farmer Jones raises chickens and donkeys. Chickens have 2 legs, and donkeys have 4 legs. If Farmer Jones owns 80 animals and the animals have a total of 232 legs, how many chickens does he own?

70. A fish tank is 50% full. Its length is twice its width and its height is twice its length. If the length is 8 inches, what is the volume of the water (in cubic inches)?

71. 3 inch long squares are cut out of the corners of a rectangle with dimensions of 11 inches by 14 inches. If the resulting net is folded to make a box without a lid, what is the volume of this box (in cubic inches)?

72. If a regular hexagon and an equilateral triangle have the same perimeter, what is the ratio of the area of the hexagon to that of the triangle? Express your answer as a common fraction.

73. A right circular cone has a radius of length r centimeters. It has a height of 6 centimeters and the length of its slant height is $r + 2$ centimeters. What is that value of r?

74. Mike's bicycle has a front wheel of diameter 2 feet and a rear wheel of radius 9 inches. When Mike rides his bike to the grocery store, his front wheel goes through 300 revolutions. How many revolutions does his rear wheel go through?

75. What are the rightmost two digits of 5^{555}?

76. Jane the macaw is flying through the rainforest. She spots 100 frogs on her 2 hour flight. If she never saw two frogs within 30 seconds or shorter, what is the longest time she could have flown without seeing a frog?

77. Inside a circle are three smaller equal-sized circles of radius 6, where each is tangent to the outer circle and to the other two small circles. What is the radius of the large circle? Express your answer in the form $a + b\sqrt{c}$, where a, b, and c are integers, and c isn't divisible by the square of any prime.

78. If the radius of a cylinder is increased by 100%, how much should the height be decreased to maintain the same volume?

79. Two sets, A and B, are the same size. Their intersection consists of 100 elements, and their union has size 1000. Find the number of elements in set A.

80. Jim has 27 white 1×1 cubes. He decides to paint a single face of 10 cubes, 2 adjacent faces of 6 other cubes, and 2 opposite faces of 6 more cubes. When he assembles these cubes into a single 3×3 cube, what is the maximum fraction of its outside surface that will be painted?

81. A trapezoid has parallel bases having lengths of 3 units and x units. The height of the trapezoid is 6 units. If the area of the trapezoid is 30 square units, compute the value of x.

82. What is the area of a triangle with vertices at $(3, 4)$, $(7, 9)$, and $(18, 8)$?

83. A robot is standing on a conveyor belt which moves to the left at a constant rate of 5 feet per second. The robot is programmed to walk to the right at 10 feet per second. If the conveyor belt is 110 feet long and the robot starts on the farthest left side of the conveyor belt, how many seconds will it take for the robot reach to the other side of the conveyor belt?

Serious Challenges

1. Let $P(x)$ be a polynomial of degree 4. If $P(0) = \frac{1}{2}$, $P(1) = \frac{2}{3}$, $P(2) = \frac{3}{4}$, $P(3) = \frac{4}{5}$, and $P(4) = \frac{5}{6}$, compute $P(5)$.

2. How many ways are there to distribute 12 pieces of candy to 5 children? Not every child has to get candy, but all the candy must be distributed.

3. ABC is an acute triangle. Let O and H denote the circumcenter and orthocenter of the triangle respectively. If $\angle BAH = 35$ degrees, compute the value of $\angle CAO$ (in degrees) (Note: The circumcenter of a triangle is the center of the circle that circumscibes the triangle. The orthocenter of a triangle is the point of intersection of the triangle's three altitudes.)

4. Compute the last two digits of 15^{99}.

5. Compute the last two digits of $9^{3^{7^{6^5}}}$.

6. Given that a and b are integers such that $0 < a < b$ and $2ab + 3a + 7b = 100$, compute a.

7. If $a^2 + 2a + b^2 + 4b + c^2 + 6c = -14$, compute $a^2 + b^2 + c^2$.

8. Yvone and Zach are playing a game. They each roll 2 dice. Whoever has the higher sum wins. If the sums are the same, Yvone wins. What is the probability that Yvone will win?

9. A point P is chosen in the interior of square $ABCD$ such that $AP = 5, BP = 1$, and $DP = 7$. Compute the area of triangle ACP.

10. If $x^3 + 8x^2 + 6x + 3 = 0$ has roots a, b, c, compute $a^2 + b^2 + c^2$.

11. Given $x + y + z = 1$, $x^2 + y^2 + z^2 = 89$, and $x^2yz + xy^2z + xyz^2 = -84$, compute x, y, and z, if $x \leq y \leq z$.

12. Let N be an integer with 33 digits, all of which are 1's except for the 17th digit. Find the value of the 17th digit if N is divisible by 13.

13. What is the minimum value of $ab + bc + ca$ if $abc = 1$?

14. There are two barrels that look identical from the outside. Barrel A contains 6 red balls and 6 green balls, while barrel B contains 10 red balls and 2 green balls. All of the balls are identical, besides their color. Mike randomly chooses one of the barrels, and randomly draws a ball. If the ball is red, what is the probability that he picked barrel A?

15. A circle is inscribed in isosceles triangle ABC (with $AB = AC$) and is tangent to sides BC, CA, and AB at P, Q, and R, respectively. Triangle PQR is drawn. Point X is drawn on arc QR which does not contain P. If $\overline{AQ} = 3$ and $\overline{AB} = 9$, what is the value of $\dfrac{\overline{XP}}{\overline{XQ} + \overline{XR}}$?

16. Given a cubic with nonnegative roots p, q, and r, if $(p + q + r)^2 = 49$, $p^2 + q^2 + r^2 = 25$, and $pqr = 0$, what is $(x - p)(x - q)(x - r)$?

17. If the roots of the polynomial $x^4 - 5x^3 - 7x^2 + 29x + 30$ are a, b, c, and d, what is the value of $(a + 1)(b + 1)(c + 1)(d + 1)$?

Chapter 6

Solutions

Algebra
Warm-up Problems

1. When a turtle wants to cross a river, he has to pay $4. If he crosses the river 5 times on Monday, how much did he pay on Monday?

   Solution: We find that $\$4 \cdot 5 = \boxed{\$20}$.

2. Serge has a crush on Pam and decides to buy her 5 yards of yarn (and some roses). However, the yarn can only be bought in inches. How many inches of yarn does he need to buy?

   Solution: Serge will buy $5 \text{ yards} \cdot \dfrac{3 \text{ feet}}{1 \text{ yard}} \cdot \dfrac{12 \text{ inches}}{1 \text{ foot}} = \boxed{180}$ inches of yarn.

3. If a bottle of sunscreen costs $8 and you have a coupon that gives a 25% discount, how many dollars do you pay if you use the coupon?

   Solution: Since the coupon provides a 25% discount, you pay 75% of the original price of the sunscreen. This results in $0.75 \cdot 8 = \boxed{6}$ dollars.

4. Allison took a series of tests during her freshman year. Her test scores were 92, 85, 87 and 100. However, instead of 100, her test score should have been a 98.5. What is the new median? Express your answer as a decimal rounded to the nearest tenth.

Solution: We order the new set of scores from least to greatest to find the median: $\{85, 87, 92, 98.5\}$. For a set of four numbers, the median is the mean of the two middle elements, or $\dfrac{87 + 92}{2} = \boxed{89.5}$.

5. How many minutes are there in 20% of one day?

   Solution: One day contains 24 hours, and each hour is 60 minutes. Thus, there are 24×60 minutes in a day. Twenty percent, or one-fifth, of that is $\dfrac{24 \times 60}{5} = \boxed{288}$ minutes.

6. A fly eats 6 spiders every century. On average, how many spiders does the fly eat every year? Express your answer as a decimal rounded to the nearest hundredth.

   Solution: Since a century is 100 years, the fly eats $\frac{6 \text{ spiders}}{100 \text{ years}} = \boxed{0.06}$ spiders per year.

7. Emily built a toothpick bridge that was strong enough to hold 15 pounds. How many $2\frac{1}{2}$ pound rats can sit on the bridge without it breaking?

   Solution: The total number of $2\frac{1}{2}$ pound rats that weigh a total of 15 pounds is $\frac{15}{2\frac{1}{2}} = 6$ rats. So, the bridge can hold $\boxed{6}$ rats.

8. If a dog house can hold up to 5 dogs, how many dogs can 25 dog houses hold?

   Solution: 25 dog houses can hold $\frac{5 \text{ dogs}}{1 \text{ dog house}} \times 25$ dog houses $= \boxed{125}$ dogs.

9. If a frond grows at a rate of 2.5 centimeters per hour, how tall, in centimeters, is the fern after 2 days?

Solution: There are 24 hours in a day, so the height after 2 days is

$$\left(\frac{2.5 \text{ centimeters}}{1 \text{ hour}}\right)\left(\frac{24 \text{ hours}}{1 \text{ day}}\right)(2 \text{ days}) = \boxed{120} \text{ centimeters.}$$

10. If a bag of rice at 99 Ranch Market is $200, then how many dollars do 20 bags cost?

Solution: The 20 bags cost $\left(\dfrac{\$200}{1 \text{ bag}}\right)(20 \text{ bags}) = \boxed{\$4000}$.

11. You have 100 coins each worth the same amount of money. The total amount of money is $100. How much is each coin worth?

Solution: Each coin is worth $\frac{\$100}{100 \text{ coins}} = \boxed{\$1}$ per coin.

12. Let us define a function $F(x)$ such that $F(x) = 0$ only when $x = 0$ or 1. What is $F(F(1))$?

Solution: Since $F(1) = 0$, the expression becomes $F(0)$ which equals $\boxed{0}$.

Additional Problems

1. If $3x + 4y = 13$ and $4x + 3y = 22$, compute $x + y$.

   Solution: Adding these two equations, $7x + 7y = 35$ and dividing by 7 yields an answer of $\boxed{5}$.

2. Given lines $3x + 4y = 12$, $6x + 8y = 17$ and $x + y = 13$, how many points (x, y) lie on exactly two of these lines?

   Solution: Notice that the first two lines are parallel and the third line intersects both, giving $\boxed{2}$ points.

3. Victor is sharing his baguettes. If he has a supply of nine hundred baguettes, and gives out 1 during the first minute, 3 during the second minute, 5 during the third minute, and so on, how many minutes will it take for him to run out of baguettes?

   Solution: Using the fact that $1 + 3 + \cdots + (2n - 1) = n^2$, we see that since Victor has $900 = 30^2$ baguettes, it will take him $\boxed{30}$ minutes.

4. Eight students graduate from Orange County Math Circle. 17 more enter from overseas due to unsatisfactory teaching. Another 12 leave since they were assigned too much homework. There are now 26 members. How many people were in Orange County Math Circle originally?

   Solution: If there were N people in Orange County Math Circle originally, then afterwards, there are $N - 8 + 17 - 12$ in Orange County Math Circle, which must equal 26. So, $26 = N - 8 + 17 - 12 = N - 3$, and $N = \boxed{29}$ people.

5. In Calculus class, tests are worth 85% of the grade and the final is worth 15%. If William has an 86% in the test category, what grade does William need on his final to maintain a 70% in the class?

   Solution: William's total grade in percent can be calculated by $G_{Total} = 0.85 G_{Test} + 0.15 G_{Final}$, where G_{Test} is the test grade in percent and G_{Final} is the lowest percent grade he needs on the final. So, we have $70 = 0.85(86) + 0.15(G_{Final})$. So, $G_{Final} = \frac{70 - 0.85(86)}{0.15} = \frac{70 - 73.1}{0.15} = -\frac{31}{15}$ percent. Since the lowest percent grade he can receive is less than 0, he needs a $\boxed{0}$ on the final to get at least a 70% for his total grade.

6. If you walk for 45 minutes at a rate of 4 miles per hour and then run for 30 minutes at a rate of 10 mph, how many miles will you have gone at the end of the one hour and 15 minutes?

   Solution: 45 minutes is $\frac{3}{4}$ of an hour, and 30 minutes is $\frac{1}{2}$ of an hour, so while walking, you travel $\frac{3}{4}(4) = 3$ miles, and while running, you travel $\frac{1}{2}(10) = 5$ miles. So, in total, you travel $3 + 5 = \boxed{8}$ miles in one hour and 15 minutes.

7. A shirt is on sale for \$22. The shirt was discounted by 12%. What was the original price of the shirt in dollars?

   Solution: If the shirt was discounted by 12%, the discounted price is 88% of the original price. So, if the original price is P, then $\$22 = \frac{88}{100}P = \frac{22}{25}P$. So, $P = \frac{25}{22}(\$22) = \boxed{\$25}$.

8. Define a function $a <> b = a^b + b^a$. What is $(3 <> 2)(1 <> -1)$?

   Solution: $(3 <> 2)(1 <> -1) = (3^2 + 2^3)(1^{-1} + (-1)^1) = (9 + 8)(1 + (-1)) = (17)(0) = \boxed{0}$.

9. Marx is walking down a hallway when he hears a fluffy unicorn behind him. Both Marx and the unicorn begin running at the same time. Marx runs at 3 meters per second while the unicorn gallops at 10 meters per second. If Marx was trampled by the unicorn after 10 seconds, how far, in meters, was the unicorn behind Marx when Marx heard her?

   Solution: In 10 seconds, Marx runs $3 \times 10 = 30$ meters, and the unicorn gallops $10 \times 10 = 100$ meters. So, if Marx gets trampled after 10 seconds, the unicorn started off $100 - 30 = \boxed{70}$ meters behind Marx.

10. A sequence is defined as follows: $a_1 = 1$, $a_2 = 0$ and $a_k = a_{k-2} + a_{k-1}$ for $k \leq 3$. What is a_5?

    Solution: $a_3 = a_2 + a_1 = 0 + 1 = 1$, $a_4 = a_3 + a_2 = 1 + 0 = 1$, and $a_5 = a_4 + a_3 = 1 + 1 = \boxed{2}$.

11. If 5 times a number is 4, what is 200 times the reciprocal of the number?

    Solution: Let the number be N. So, $5N = 4$, and $N = \frac{4}{5}$. Therefore the reciprocal of N is $\frac{1}{N} = \frac{5}{4}$, and 200 times that is $\frac{5 \cdot 200}{4} = \frac{1000}{4} = \boxed{250}$.

12. Michelle has a pet dog weighing 20 pounds. It grows 0.7 pounds every week. It loses 0.1 pounds every day. How many pounds does it weigh after 7 weeks?

    Solution: If the dog loses 0.1 pounds every day, then it loses $0.1 \cdot 7 = 0.7$ pounds every week. Since it also gains 0.7 pounds every week, it doesn't gain or lose any pounds in 7 weeks. Thus, by the end of the week, it still weighs0 $\boxed{7}$ pounds.

13. If the average ACT score at University High School is 36 out of a group of 50 students and the average ACT score at Corona Del Mar is 39 out of a group of 100 students, what is the average ACT score of both schools?

    Solution: An average score of 36 for a group of 50 students means that the sum of all of the scores of those 50 students is $36 \cdot 50 = 1800$. An average score of 39 for a group of 100 students means that the sum of all the scores of those 100 students is $39 \cdot 100 = 3900$. So, the sum of the scores of all 150 of the students is $1800 + 3900 = 57000$, and the average score is $\frac{57000}{150} = \boxed{38}$.

14. A CAM costs \$2000 while a CP costs \$30,000. How many CAMs can you buy for the price of a CP?

    Solution: You can buy $\frac{\$30,000}{\$2000} = \boxed{15}$ CAMs for the price of a CP.

15. Jessica is taking a Spanish class at school. This semester, she has received test scores of 90, 84, 96, and 98. She wishes to

receive an A this semester, which requires a test average of 93. Compute the minimum score that she must receive for her last test in order to receive an A.

Solution: Since Jessica wants a minimum test average of 93, she needs to receive a total test score on these 5 tests of at least $93 \cdot 5 = 465$. Thus, the score that she must receive for her last test is $465 - (90 + 84 + 96 + 98) = \boxed{97}$.

16. Two people are running toward each other at 5 miles per hour in opposite directions on adjacent sidewalks. A bird decides to have some fun and flies between the two people at 18 miles per hour until the runners meet. It begins flying when the runners are 20 miles apart. What is the total distance the bird will fly?

Solution: The bird will fly until the runners meet, or $20 \div (2 \cdot 5) = 2$ hours. Thus, the bird will fly for a total distance of $2 \cdot 18 = \boxed{36}$ miles.

17. Jon begins eating a box of 50 cookies at a rate of 2 cookies per minute. Wishing to join in the fun after watching Jon eat for 5 minutes, Allie shares Jon's cookies at a rate of 3 cookies per minute. Jon is feeling a little full, so when Allie begins eating cookies with him, he slows down to a rate of 1 cookie per minute. How many cookies has Allie eaten when they finish the box?

Solution: When Allie starts eating, there are $50 - 2 \cdot 5 = 40$ cookies left. Together, Allie and Jon eat at a rate of $1 + 3 = 4$ cookies per minute. Thus, it takes them $40 \div 4 = 10$ more minutes to finish the box. In the end, Allie has eaten $10 \cdot 3 = \boxed{30}$ cookies.

18. At 2:30 PM what is the smaller angle is formed by the hour and minute hand?

Solution: The minute hand of a clock is at an angle of $6m$ from the vertical, so at 2 : 30 PM, the minute hand is at $180°$ from the vertical. The hour hand is at $30h + \frac{1}{2}m$, so the hour hand is at $75°$ from the vertical. So, the smaller angle between the hands is $180° - 75° = \boxed{105°}$.

19. If $a = -1$, compute the largest number in the set

$$\left\{-3a, 4a, \frac{24}{a}, a^2, 1\right\}.$$

Solution: Plugging in $a = -1$, the set becomes

$$\left\{-3(-1), 4(-1), \frac{24}{-1}, (-1)^2, 1\right\}$$

$$= \{3, -4, -24, 1, 1\}$$

So, the largest number is $\boxed{3}$.

20. Dorky buys a shirt for $20 dollars and is charged an additional 6.5% sales tax. Lucky buys the same shirt but is charged an additional 6% sales tax. How many more cents does Dorky pay than Lucky?

Solution: A 6.5% sales tax equals $\frac{6.5}{100}(\$20) = \$1.30$, and a 6% sales tax equals $\frac{6}{100}(\$20) = \$1.20$. So, Dorky has to pay $\$1.30 - \$1.20 = \boxed{10}$ cents more.

21. Mr. Frahit receives a 10% raise every year. His salary after four such raises has gone up by what percent?

Solution: A raise of 10% means that his salary gets multiplied by $\frac{11}{10}$. Four such raises means that his salary is multiplied by $11/10$ four times, which is equal to multiplying it by $\left(\frac{11}{10}\right)^4 = \frac{14641}{10000}$. This means his new salary is $(100 \cdot \frac{14641}{10000})\% = 146.41\%$ of his original salary, which is an increase of $\boxed{46.41\%}$.

22. If the French Army retreats 3 times faster than than the British Army, and the British Army retreats 9 times faster than the German Army, how fast (in meters per second) do the French run if the Germans retreat at a rate of 8 meters per second from Pakistan?

Solution: The British retreat 9 times faster than the Germans, so they retreat at $9 \cdot 8 = 72$ meters per second. The French retreat 3 times faster than that, so they retreat at $72 \cdot 3 = \boxed{216}$ meters per second.

23. This year, $700,000$ people took the AMC. 5% of these passed into AIME. 10% of the people who take the AIME can take the USAMO. 0.2% of USAMO takers pass into IMO. How many people passed into the IMO this year?

Solution: Five percent of $700,000$ people is $(0.05)700,000 = 35000$ people who passed into AIME. Ten percent of 35000 people is $(0.10)35000 = 3500$ people who passed into US-AMO. 0.2% of 3500 people is $(0.002)3500 = \boxed{7}$ people who passed into the IMO.

24. Sale prices at the OCMC Outlet Store are 50% below original prices. On Saturdays an additional discount of 20% off the sale price is given. What is the Saturday price (in dollars) of a coat whose original price is $180?

Solution: A regular sale price would be 50% of the original price, or $\frac{1}{2}(\$180) = \$90$. An additional discount 20% means that the Saturday price is 80% of the regular sale price, or $\frac{4}{5}(\$90) = \boxed{\$72}$.

25. Jane went to the zoo and saw that there were a lot of visitors that day. Since there was a long wait before she could enter the zoo, she counted tires, sedans, and motorcycles in the parking lot. If she counted a total of 100 vehicles and 340 tires, how many motorcycles did she see?

147

Solution: Let $c =$ the number of sedans, and $m =$ the number of motorcycles. Then we find that $c + m = 100$, $4c + 2m = 340$. Thus, we find that $c = 70$, and $m = \boxed{30}$.

26. Mary is shopping at her favorite mall. In the first store she shops in, Mary spends 20% of her money. Then, she spends 25% of her remaining money on fancy pens and pencils. Before she goes home, Mary eats lunch, saving 80% of the money she had left. When she arrives at her house, she finds that she has \$24 remaining in her purse. How much money did she bring to the mall?

Solution: Let the original amount of money that Mary had be x. After the first store, Mary has $0.8x$ left, and then spends 25% of that money. We see that before she eats lunch, she has $75\% \cdot 0.8x = 0.6x$ left. When eating lunch, she leaves 80% of her remaining money, so the amount of money she counts when arriving home is equivalent to $80\% \cdot 0.6x = 0.48x = 24$. Thus, we find that $x = \boxed{50}$.

27. Let $a \bowtie b = \dfrac{2a + b}{a^2 + b^2}$. Compute $3 \bowtie (2 \bowtie 1)$.

Solution: This is a simple plug and chug. We evaluate $2 \bowtie 1$ first, which evaluates to $\dfrac{2 \cdot 2 + 1}{2^2 + 1^2} = 1$. Then $3 \bowtie 1 = \dfrac{2 \cdot 3 + 1}{3^2 + 1^2} = \boxed{\dfrac{7}{10}}$.

28. A poodle usually drinks water at a rate of 1 liter per 7 minutes. Since the poodle is very thirsty, its drinking rate triples. If we leave six-sevenths of a liter out for the poodle, in how many minutes will it finish?

Solution: The very thirsty poodle will now drink water at a rate of 3 liters per 7 minutes. Thus, six-sevenths of a liter will last $\dfrac{6}{7} \div \dfrac{3}{7} = \boxed{2}$ minutes.

29. I am a number between 1 and 100. You get the same number if you multiply me by 4 or add 198 to me. What number am I?

Solution: Let the number we must find be x. Conclude that $4 \cdot x = x + 198$, so $3x = 198$. Thus, the number is $\boxed{66}$.

30. A dog runs toward a ball 30 feet away. When the dog is halfway there, how much farther does he have to run to get the ball and get back to his starting point (in feet)?

Solution: The distance the dog must still run to reach the ball is $\frac{30}{2} = 15$ feet. Then, he must run 30 feet to return to his starting point. This results in a total of $15 + 30 = \boxed{45}$ feet.

31. If the beaver slaps his tail on water 4 times over 5 seconds, then pauses for 5 seconds, then repeats the cycle, how many times does he slap his tail on the water in one minute?

Solution: We see that for every ten seconds that pass, the beaver slaps his tail on the water 4 times. Thus, he slaps $\frac{4}{10} \cdot 60 = \boxed{24}$ times in one minute, which is equivalent to 60 seconds.

32. The product of all 4 sides of a square is 1296. What is the sum of all four of the square's sides?

Solution: Let the length of a side of the square be x. Then $x^4 = 1296$, so $x = 6$. We find that the sum of all four of the square's sides is simply $4x = 4 \cdot 6 = \boxed{24}$.

33. John has some cookies. He gives half of them to his siblings, one-third of them to his parents, and one-twelfth to his friend. He eats the only cookie left. How many cookies did he start with?

Solution: Let the number of cookies John has originally be x. The number of cookies he has left after giving them to his siblings, parents, and friend is $(1 - \frac{1}{2} - \frac{1}{3} - \frac{1}{12}) \cdot x = \frac{1}{12}x$. We find that $\frac{1}{12}x = 1$, yielding $x = \boxed{12}$ cookies.

34. If Josh took the square root of a positive number, tripled that number, and then added 8, he got the number 20. But instead of taking the square root of the number, he should have squared it. If Josh had squared the number, what would his result be?

    Solution: We work backward from 20 to find Josh's positive number, say j. We find that $j = \{(20-8) \div 3\}^2 = 16$. To find the result, we follow the same process, but instead squaring the number instead of finding the square root. This results in $16^2 \cdot 3 + 8 = \boxed{776}$.

35. The sum of 10 consecutive integers, starting with 11, equals the sum of 5 consecutive integers, starting with what number?

    Solution: We first find the sum of the 10 consecutive integers. We see that the last of these integers is $11 + 10 - 1 = 20$, so the sum of the 10 integers is $\frac{10}{2}(11 + 20) = 155$. Let the desired first integer be x. We see that the last of the 5 integers is $x + 4$, so the sum of the 5 integers is $\frac{5}{2}(x + x + 4) = 5(x + 2) = 155$. Thus, $x = \boxed{29}$.

36. A room is one-half full of people. After 20 people leave, the room is one-third full. How many people would be in the room if it were full?

    Solution: Let the capacity of the room be x. We see that the situation becomes $\frac{1}{2}x - 20 = \frac{1}{3}x$, hence $\frac{1}{6}x = 20 \implies x = \boxed{120}$.

37. A cell phone battery is at 40% and will run out in 36 minutes. How long (in minutes) will the battery last if it is at 90%?

    Solution: We see that if x is the value we would like to find, the situation becomes $\dfrac{40\%}{36} = \dfrac{90\%}{x}$. Therefore, $x = \boxed{81}$.

150

38. My age is 9 less than 2 times my sister's age. In 4 years, the sum of our ages will be twice my brother's current age, which is 5 more than my sister's current age. What is my age?

    Solution: Let the sister's age be x. We see that my age is $2x-9$, while my brother's age is $x+5$. In 4 years, the situation will become $(2x-9+4)+(x+4) = 2 \cdot (x+5)$, which simplifies to $x = 11$. Therefore, my age is $2 \cdot 11 - 9 = \boxed{13}$.

39. The sum of 9 consecutive integers is 90. What is the largest of these integers?

    Solution: Let the largest of the 9 consecutive integers be x. We see that the least of the integers is $x - 9 + 1 = x - 8$, so the sum of the 9 consecutive integers is $\frac{9}{2}(x - 8 + x) = 9(x - 4) = 90$. $x = \boxed{14}$.

40. If $x^2 - y^2 = 96$ and $x + y = 16$, then what is $x - y$?

    Solution: Recall that $x^2 - y^2 = (x + y)(x - y) = 96$. Therefore, $16(x + y) = 96$, and $x - y = \boxed{6}$.

41. Michelle is taking a geometry class in which an average of 85 or above on tests is needed to pass. Her scores so far have been 98, 86, 87, and 64. What is the lowest score she can get on her fifth test and still pass?

    Solution: The lowest sum of the five tests that Michelle can receive to pass is $85 \cdot 5 = 425$ points. We see that the lowest score she can get on her fifth test is $425 - (98 + 86 + 87 + 64) = \boxed{90}$.

42. It takes 5 men to build 3 cars in 8 days. How many days will it take 2 men to build 6 cars?

    Solution: We see that building one car takes $8 \cdot 5 \div 3 = \frac{40}{3}$ man-days. To build 6 cars using only 2 men, this will then take $\frac{40}{3} \cdot 6 \div 2 = \boxed{40}$ days.

151

43. A bike and a car are 210 miles apart. The bike is going at 10 miles per hour in the direction of the car and the car is driving at 60 miles per hour in the direction of the bike. How many hours will it take for the two vehicles to reach each other?

Solution: We see that since the bike is going at 10 miles per hour toward the car and the car is driving at 60 miles per hour toward the bike, the two vehicles are moving together at a rate of 70 miles per hour. Thus, the two vehicles will take $210 \div 70 = \boxed{3}$ hours to reach each other.

44. A function $f(x)$ is defined by $f(x) = f(x-1) + f(x-2)$. If $f(1) = 2$ and $f(5) = 25$, find $f(4)$.

Solution: Let $f(2) = x$. Then, $f(3) = x + 2$, $f(4) = x + (x+2) = 2x + 2$, and $f(5) = (x+2) + (2x+2) = 3x + 4 = 25$. So, $x = 7$, and $f(4) = 2(7) + 2 = \boxed{16}$.

45. If a driver drives five miles at 50 miles per hour, how fast should he drive for the next 3 miles so that his average speed for the whole trip is 60 miles per hour?

Solution: The average speed for the whole trip is calculated by dividing the total distance by the time taken. So, we have $60 = \frac{8}{T}$, and $T = \frac{2}{15}$ hours, or 8 minutes, where T is the total time it takes to drive the 8 miles. If V is the speed for the last 3 miles, then $T = \frac{5}{50} + \frac{3}{V}$, so $\frac{3}{V} = \frac{2}{15} - \frac{1}{10} = \frac{1}{30}$, and $V = \boxed{90}$ miles per hour.

46. Bob has taken 6 tests so far in his math class, each worth a total of 100 points. His grade in math class depends solely on his test scores, and is currently exactly 88%. He is going to take the final next week, and wants to know if he can bring his grade up to 90%. The final is worth 150 points. What grade in percent must Bob get on his final to bring his grade up to 90%?

Solution: Bob's 6 tests are worth 600 points total. He currently has an 88%, which means he has $\frac{88}{100}(600) = 528$ points out of 600. After the final, the total number of possible points will be brought up to 750. To get a 90%, he needs $\frac{9}{10}(750) = 675$ points. That means he needs to get $675 - 528 = 147$ points out of 150 on the final, which is $\boxed{98\%}$.

47. How many ways can $5 be formed with quarters and pennies?

Solution: Using 0 quarters would require 500 pennies and using 20 quarters would require 0 pennies. In addition, for any amount of quarters between 0 and 20 there exists an amount of pennies needed to form $5. Thus, our answer is simply the number of integers between 0 and 20 inclusive, or $20 - 0 + 1 = \boxed{21}$.

48. If $x^2 + y^2 = 12$ where x and y are real numbers, what is the maximum value of xy?

Solution: By AM-GM, $\frac{x^2+y^2}{2} \geq \sqrt{x^2y^2}$, or $x^2 + y^2 \geq 2xy$ (alternatively, notice that $(x-y)^2 \geq 0$ and rearrange). Thus, $2xy \leq 12$ and $xy \leq \boxed{6}$.

49. Adam says that his favorite number is a positive number x such that $x = \cfrac{1}{x+\cfrac{1}{x+\cfrac{1}{x+\cdots}}}$. Compute Adam's favorite number.

Solution: Notice that $x = \frac{1}{x+\cdots}$ where $\cdots = x$. Thus, $x = \frac{1}{x+x}$, or $x = \frac{1}{2x}$ so $x^2 = \frac{1}{2}$ and $x = \boxed{\dfrac{\sqrt{2}}{2}}$.

50. If the answer to this problem is x, what is $x^2 + 3x + 1$?

Solution: Since the answer to the problem is both x and $x^2 + 3x + 1$, we have $x = x^2 + 3x + 1$, so $x^2 + 2x + 1 = 0$, and $(x + 1)^2 = 0$, so $x + 1 = 0$, and $x = -1$. Therefore, the answer to the problem is $\boxed{-1}$.

51. At the beginning of a trip, the mileage odometer read $56,200$ miles. The driver filled the gas tank with 6 gallons of gasoline. During the trip, the driver filled his tank again with 12 gallons of gasoline when the odometer read $56,560$. At the end of the trip, the driver filled his tank again with 20 gallons of gasoline. The odometer read $57,060$. To the nearest tenth, what was the car's average miles-per-gallon for the entire trip?

Solution: The driver started the trip with a full tank and had to fill it twice, once with 12 gallons and then with 20 gallons. So, in total, the trip used up 32 gallons of gasoline. Since the odometer started at $56,200$ miles and read $57,060$ miles at the end of the trip, the driver drove $57,060 - 56,200 = 860$ miles. Thus, the average miles-per-gallon was $\frac{860 \text{ miles}}{32 \text{ gallons}} = \boxed{26.9}$ mpg.

52. Michelle wants to go to Pakistan to visit her dog. She goes online to buy tickets, but her computer has a mysterious virus that turns numbers into algebraic expressions. Pakistan Airlines shows that her ticket costs $\frac{x^2 - x - 6}{x+2}$ dollars where $x = 1340$. How much money does her ticket cost (in dollars)?

Solution: $x^2 - x - 6 = (x-3)(x+2)$, so $\frac{x^2-x-6}{x+2} = x - 3 = \boxed{1337}$ dollars.

53. What is the sum of the first 13 terms of the following arithmetic sequence: $30, 26, 22, 18, ...$?

Solution: The difference is $30 - 26 = 4$, so the first 13 terms are $30, 26, 22, 18, 14, 10, 6, 2, -2, -6, -10, -14, -18$. The sum of the first 13 terms must be

$$30 + 26 + 22 + 18 + 14 + 10 + 6 + 2 - 2 - 6 - 10 - 14 - 18 - 22$$

$$= 30 + 26 + 22 + (18 - 18) + (14 - 14) + (10 - 10) + (6 - 6) + (2 - 2)$$

$$= 30 + 26 + 22 = \boxed{78}$$

54. Let x be an integer such that $x^2 - 8x + 11 < -4$. Compute x.

Solution: If $x^2 - 8x + 11 < -4$, then $x^2 - 8x + 15 < 0$, so $(x - 3)(x - 5) < 0$. If $x = 3$ or 5, then $(x - 3)(x - 5) = 0$. If $x < 3$, then both $(x - 3)$ and $(x - 5) < 0$, so $(x - 3)(x - 5) > 0$. If $x > 5$, then both $(x - 3)$ and $(x - 5) > 0$, so $(x - 3)(x - 5) > 0$. So for $(x - 3)(x - 5) < 0$, $3 < x < 5$, so $x = \boxed{4}$.

55. Jenny buys 10 cupcakes from Whipped Cream Cupcakes. Her friend Jessica buys 14 cupcakes at the store the next day. All of the cupcakes in the store cost \$2. The cashier tells both of them that Whipped Cream Cupcakes offers a sale such that any purchase of 18 or more cupcakes gets 4 free cupcakes. How much could Jenny and Jessica have saved on their cupcakes if they had bought them together?

Solution: Jenny spends $2 \cdot 10 = 20$ dollars on cupcakes, while Jessica spends $2 \cdot 14 = 28$ dollars on her cupcakes. When they buy them together, they want to have a total of 24 cupcakes, so they buy $24 - 4 = 20$ cupcakes, getting 4 free. This costs $20 \cdot 2 = 40$ dollars. Thus, they would have saved $20 + 28 - 40 = \boxed{8}$ dollars.

56. What is the measure in degrees of the acute angle formed by the hour and minute hands of a clock at 2 : 20 PM?

Solution: Let the direction of the 12 serve as $0°$. In 1 hour, the minute hand traverses an entire circle, so for each minute, the minute hand traverses $360° \div 60 = 6°$. The hour hand traverses an entire circle only after 12 hours, so each hour is equivalent to $360° \div 12 = 30°$, while each minute is $30° \div 60 = 0.5°$. The minute hand is at 20 minutes, or $20 \cdot 6° = 120°$. The hour hand is at 2 hours and 20 minutes, or $2 \cdot 30° + 20 \cdot 0.5° = 70°$. Thus, the acute angle formed by the two hands is $120° - 70° = \boxed{50°}$.

57. Compute the number of integers x that will make $\dfrac{2x^2 - 4x + 15}{x - 5}$ an integer.

Solution: Using long division, we find that $\dfrac{2x^2 - 4x + 15}{x - 5} = 2x + 6 + \dfrac{45}{x - 5}$. Since x is an integer, we know that $2x + 6$ is as well, so we just need to ensure that $\dfrac{45}{x - 5}$ is an integer. Thus, $x - 5$ must be a divisor of 45. Finding the prime factorization of 45, we see that $45 = 3^2 \cdot 5$, so it has $(2+1)(1+1) = 6$ positive factors: $1, 3, 5, 9, 15, 45$. The question asks for all integers, we also include the negative factors $-1, -3, -5, -9, -15, -45$, bringing us to a total of $\boxed{12}$ different integers.

58. If it takes 12 men 12 hours to build 12 robots, how much time does it take 6 men to build 6 robots?

Solution: 12 men work at the pace of 12 robots per 12 hours, which is equal to 1 robot per hour. Hence, 1 man works at the pace of 1 robot per 12 hours, and 6 men work at the pace of 1 robot per 2 hours. Hence, building 6 robots requires $\boxed{12}$ hours.

59. In the book <u>Moles</u> by Souis Lachar, moles must dig holes shaped like a cube with length 6 feet. Three moles can dig 9 cubic feet in 9 minutes. How many moles does it take to finish a complete hole in 27 minutes?

Solution: Since the volume of a cube is its side length cubed, the volume of each hole is $6^3 = 216$ cubic feet. 3 moles can dig 9 cubic feet in 9 minutes, 3 moles can dig 27 cubic feet in 27 minutes. Therefore, you need $\frac{216}{27}(3) = 8(3) = \boxed{24}$ moles to dig 216 cubic feet in 27 minutes.

60. If a woodchuck chucks 10 chucks of wood in 3 hours, how many chucks of wood can 3 woodchucks chuck in 1 hour?

Solution: 3 woodchucks chucking for 3 hours chuck 3 times as much wood as 1 woodchuck for 3 hours. 3 woodchucks for

1 hour chuck a third as much wood as 3 woodchucks for 3 hours, so 3 woodchucks chucking for 1 hour chuck as much as 1 woodchuck chucking for 3 hours, which is $\boxed{10}$ chucks of wood.

61. Compute the value of $1 + \cfrac{1}{2 + \cfrac{1}{1 + \cfrac{1}{2 + \cfrac{1}{1 + \cdots}}}}$.

Solution: Let the value be x. Since this is an infinite fraction, we can substitute so that the value we wish to compute becomes $1 + \cfrac{1}{2 + \cfrac{1}{x}}$, which equals x. Solving the resulting quadratic for x, we find that since the value must be positive, $x = \boxed{\dfrac{1 + \sqrt{3}}{2}}$.

62. If the cost of an apple and 2 oranges is 60 cents, the cost of an orange and 2 peaches is 36, the cost of a peach and 2 apricots is 120, and the cost of an apricot and 2 apples is 96, what is the cost of buying one of each (in cents)?

Solution: Let each of the above costs be individual transactions. Adding all four transactions up, we find that we buy three of each fruit for a total cost of $60 + 36 + 120 + 96 = 312$ cents. Thus, the cost of buying one of each is $312 \div 3 = \boxed{104}$ cents.

63. Find the sum of all perfect cubes that are less than 250.

Solution: The sum of the first n cubes is the square of the nth triangular number. Since $6^3 = 216$ is the largest perfect cube less than 250, our answer is simply the square of the 6th triangular number, which is 21. So we get $21^2 = \boxed{441}$ as our answer.

64. Let $a \wr b = \dfrac{a|b - a|}{b}$ and $a \dagger b = a \wr b + b \wr a$. What is $(4 \dagger 6) \wr (6 \dagger 4)$?

157

Solution: We could compute $(4 \dagger 6)$ and $(6 \dagger 4)$ and then substitute both into the $a \wr b$ function, but we look for an easier way. From the definition of $a \dagger b$, we see that this function is commutative, namely that $a \dagger b = b \dagger a$. Further, when we examine $a \wr b$, we see that if $a = b$, the $|b - a|$ in the numerator of the definition will cause the function to equal zero. Thus, we see that $(4 \dagger 6) \wr (6 \dagger 4)$ evaluates to $\boxed{0}$.

65. A movie theater sells adult tickets for 6 dollars and children tickets for 4 dollars. 300 people attended one showing, and the theater made \$1560. How many adults watched the movie?

Solution: Let the number of adult tickets be A and the number of children tickets be C. The situation becomes $A + C = 300$, and $6A + 4C = 1560$, which simplifies to $3A + 2C = 780$. Solving the system of equations, we find that $C = 120$, and $A = \boxed{180}$ tickets.

66. Let $a \triangle b = \frac{a+b}{ab}$ and $a \angle b = \frac{b-a}{ab}$. What is $(123 \triangle 456) + (41 \angle 456)$? Express your answer as a common fraction.

Solution: We can separate $\frac{a+b}{ab}$ into $\frac{a}{ab} + \frac{b}{ab} = \frac{1}{b} + \frac{1}{a}$. So, $a \triangle b = \frac{1}{a} + \frac{1}{b}$. Similarly, $a \angle b = \frac{b-a}{ab} = \frac{b}{ab} - \frac{a}{ab} = \frac{1}{a} - \frac{1}{b}$. So, $(123 \triangle 456) + (41 \angle 456) = \frac{1}{123} + \frac{1}{456} + \frac{1}{41} - \frac{1}{456} = \frac{1}{123} + \frac{1}{41} = \boxed{\dfrac{4}{123}}$.

67. In 6 years, I will be a third as old as my father. In 24 years, I will be half as old as him. How old am I?

Solution: Let the father's age be F and the son's age be S. Then, we have $(S + 6) = \frac{1}{3}(F + 6)$ and $(S + 24) = \frac{1}{2}(F + 24)$. So, $3S + 18 = F + 6$ and $2S + 48 = F + 24$. Therefore, $3S + 36 = F + 24 = 2S + 48$, and $S = 48 - 36 = \boxed{12}$.

Challenge Problems

1. For $x > 0$, given that $x^2 + 1/x^2 = 7$, compute $x + 1/x$.

   Solution: Note that $x^2 + 2 + 1/x^2 = 9$, and $(x + 1/x)^2 = 9$ so $x + 1/x = \boxed{3}$.

2. Susan left her house at around 8:15 PM for a walk. When she came back, she realized that on her mechanical clock, the two pointers overlapped. She hasn't been out for more than an hour. When did she come back? Round your answer to the nearest minute. Express your answer in the form Hour : Minute.

   Solution: If Susan hasn't left the house for more than an hour after 8:15 PM, the time must still be 8 and some minutes. Let the number of minutes be x. The minute hand sweeps out one full circle on the clock in 60 minutes, so each minute the minute hand sweeps out $360° \div 60 = 6°$. The hour hand sweeps out one full circle on the clock in 12 hours, so each minute the hour hand sweeps out $360° \div (12 \cdot 60) = 0.5°$, and each hour the hour hand sweeps out $360° \div 12 = 30°$. Consider the top of the clock, or the 12, as the $0°$ point. At 8:00 PM, the hour hand is at $8 \cdot 30° = 240°$. We see that for the two pointers to overlap, $240 + 0.5x = 6x$, producing $x \approx 44$ minutes. Thus, Susan comes back at $\boxed{8:44}$ PM.

3. Pipes A and B flow into a 1,000 liter tank while pipes C and D flow out. The tank starts out empty. Pipes A and B are opened, while C and D are closed, and the tank takes 4 hours to fill up. A and B are then closed, while C and D are opened, and the tank takes 6 hours to empty. Pipes A and B are opened, C is left open, and D is closed. The tank takes 8 hours to fill up. If pipes A, B, and C are closed, and D is opened, how long will it take for the tank to fill?

   Solution: Let A, B, C, and D represent the rate water flows through pipes A, B, C, and D in tanks filled per hour, respectively, so that A and B are positive and C and D are negative.

159

So, $A + B$ is the rate of A and B together, and $\frac{1}{A+B}$ is the number of hours it takes A and B to fill the tank together. So, we have $\frac{1}{A+B} = 4$ hours, $\frac{-1}{C+D} = 6$ hours, and $\frac{1}{A+B+C} = 8$ and want to find $\frac{-1}{D}$. So, $A + B = \frac{1}{4}$ and $C + D = -\frac{1}{6}$, so $A + B + C + D = \frac{1}{4} - \frac{1}{6} = \frac{1}{12}$. Since $A + B + C = \frac{1}{8}$, $D = (A + B + C + D) - (A + B + C) = \frac{1}{12} - \frac{1}{8} = -\frac{1}{24}$. So, pipe D would take $\frac{-1}{D} = \boxed{24}$ hours to empty the tank.

4. $f(x) = x^3 + ax^2 + bx + c$ is a cubic. If $f(-11) = f(-2) = f(2) = 48$, find an expression for $f(x)$.

Solution: Let $g(x) = f(x) - 48$. Then, $g(x)$ is also a cubic, and $g(-2) = g(2) = g(11) = 0$, so $g(x) = (x + 11)(x + 2)(x - 2) = (x + 11)(x^2 - 4) = x^3 + 11x^2 - 4x - 44$. So, $f(x) = g(x) + 48 = \boxed{x^3 + 11x^2 - 4x + 4}$.

5. If $x + \frac{1}{x} = a$ and $x^2 + \frac{1}{x^2} = a$, where x and a are real numbers, what are the possible values for a?

Solution: Squaring the first equation gives $a^2 = (x + \frac{1}{x})^2 = x^2 + \frac{1}{x^2} + 2(x)(\frac{1}{x}) = a + 2$. So, $a^2 - a - 2 = 0 = (a - 2)(a + 1)$. This gives $a = 2$ and $a = -1$. We can quickly see that $x = 1$ will give us $a = 2$, so that answer is correct. However, since x is real, x^2 and $\frac{1}{x^2}$ are nonnegative, so $x^2 + \frac{1}{x^2} = a$ must also be nonnegative. Therefore, $a = -1$ is not a possible value for a, leaving $a = \boxed{2}$ as the only solution.

6. At sunrise, two people start to walk towards each other at constant, but different, speeds. One starts from town A and goes towards town B, while the other starts from town B and goes towards town A, following the same path in the opposite direction. At noon, they pass each other. The first person reaches town B at 4:00 p.m., while the second person reaches town A at 9:00 p.m. When was sunrise that day?

Solution: Let the point that the two people meet at noon be C, the distance from A to C be D_A, and the distance from B to C be D_B. Let the speed of the first person be V_A,

160

and the speed of the second person be V_B. Let the time in hours from sunrise to noon be T. Then, since Time $= \frac{\text{Distance}}{\text{Speed}}$, $T = \frac{D_A}{V_A} = \frac{D_B}{V_B}$, and $\frac{D_A}{V_A} = 4$ and $\frac{D_B}{V_A} = 9$. We can manipulate these equations to get $\frac{T}{4} = \frac{\frac{D_A}{V_A}}{D_A V_B} = \frac{V_B}{V_A}$, and $\frac{9}{T} = \frac{\frac{D_B}{V_A}}{\frac{D_B}{V_B}} = \frac{V_B}{V_A}$. So, $\frac{9}{T} = \frac{T}{4}$, $T^2 = 36$, and $T = 6$ hours. Therefore, sunrise was at $\boxed{6 : 00 \text{ a.m.}}$

7. Paul and Rick are both painting rooms. Paul can paint a room in 3 hours, but if they work together, they can paint one room in 72 minutes. How many days would it take them to paint 43 rooms working together if Paul works for 8 hours a day and Rick works for 9 hours a day? Assume that the hours they spend working do not include travel time from one room to the next.

Solution: Let the rate at which Paul works in rooms per hour be r_P, and the rate at which Rick works be r_R. We have $r_P = \frac{1\text{room}}{3\text{hours}} = \frac{1}{3}$ rooms per hour, and $r_P + r_R = \frac{1\text{room}}{\frac{6}{5}\text{hours}} = \frac{5}{6}$ rooms per hour. So, $r_R = \frac{5}{6} - \frac{1}{3} = \frac{1}{2}$ rooms per hour. Now, let R_P be the number of rooms per day Paul paints and R_R be the number of rooms per day Rick paints. Since Paul works 8 hours and Rick works 9 hours, $R_P = 8r_P = \frac{8}{3}$ rooms per day, and $R_R = 9r_R = \frac{9}{2}$ rooms per day. In total, they paint $R_P + R_R = \frac{8}{3} + \frac{9}{2} = \frac{43}{6}$ rooms per day. Therefore, Paul and Rick need $\left\lceil \frac{43}{\frac{43}{6}} \right\rceil = \boxed{6}$ days to paint 43 rooms.

8. Compute: $\frac{1}{2 \times 4} + \frac{1}{4 \times 6} + \frac{1}{6 \times 8} + \cdots$

Solution: In summation notation, this sum is

$$\sum_{n=1}^{\infty} \frac{1}{(2n)(2n+2)}$$

We look at the partial fraction decomposition of $\frac{1}{(2n)(2n+2)}$,

161

which turns out to be

$$\frac{\frac{1}{2}}{2n} - \frac{\frac{1}{2}}{2n+2}$$

As a result,

$$\sum_{n=1}^{\infty} \frac{1}{(2n)(2n+2)} = \frac{1}{2}\left(\sum_{n=1}^{\infty} \frac{1}{2n} - \sum_{n=1}^{\infty} \frac{1}{2n+2}\right)$$

$$= \frac{1}{2}\left(\frac{1}{2} + \sum_{n=2}^{\infty} \frac{1}{2n} - \sum_{n=1}^{\infty} \frac{1}{2n+2}\right)$$

Now, the two summations are the same, so the answer $\frac{1}{2} \cdot \frac{1}{2} =$ $\boxed{1/4}$.

9. What is the minimum value of the expression $x^2 + 8y + z^2 + 25 - 6x + 15 + y^2 - 12z$? What is the value of xyz when this expression is at its minimum value?

Solution: This expression can be reduced to a sum of squares, namely $(x-3)^2 + (y+4)^2 + (z-6)^2 - 21$. This means the minimum value is $\boxed{-21}$, and is reached when $x = 3, y = -4, z = 6 \rightarrow xyz = \boxed{-72}$.

Geometry

Warm-up Problems

1. What is the area of a circle whose circumference is equal to 12π? Express your answer in terms of π.

   Solution: Since the circumference is 12π, the radius must be 6. It follows that the area of the circle is $6^2\pi = \boxed{36\pi}$.

2. If right triangle ABC has $AB = 34$ and $BC = 16$, what is the maximum possible side length of AC?

   Solution: To achieve the maximal length for AC, we use the two given sides as the legs of the right triangle and set AC as the hypotenuse. This gives us the answer of $\boxed{2\sqrt{353}}$.

3. Parallelogram ABCD has $\angle ABC = 50°$. The angle bisector of $\angle BCD$ meets side AB at a point E between A and B. What is the measure of $\angle CEA$?

   Solution: Since ABCD is a parallelogram, we know that $\angle BCD = 180° - 50° = 130°$. Then we have $\angle BCE = 65°$ since E is on the angle bisector, so we know that $\angle CEA = 65° + 50° = \boxed{115°}$ by the exterior angle theorem.

4. Compute the distance (in units) between the points $(2, -1)$ and $(8, 7)$.

   Solution: Using the distance formula gives us

   $$\sqrt{(8-2)^2 + (7-(-1))^2} = \sqrt{100} = \boxed{10}.$$

5. The side lengths of a triangle are 9 inches, 12 inches, and 15 inches. In square inches, what is the area of the triangle?

   Solution: By the Pythagorean Theorem, the triangle is a right triangle, so the area is $\frac{9 \cdot 12}{2} = \boxed{54}$ square inches.

6. A 40-degree sector is cut out of a circle with radius 90 feet. What is the perimeter of the sector (in feet)? Express your answer in terms of π.

Solution: Since the central angle of the sector is 40 degrees, the length of the arc is $\frac{40}{360} \cdot 2 \cdot 90 \cdot \pi = 20\pi$. Adding on the straight parts of the sectors (radii of the original circle), we see that the perimeter is $2 \cdot 90 + 20\pi = \boxed{180 + 20\pi}$.

7. Johnny wants to plant flowers in a 3 feet by 3 feet square garden and place a stone walkway 3 feet wide around the garden (The outer edges of the walkway also form a square). If the flowers cost \$4 per square foot and the walkway costs \$6 per square foot, how much will Johnny have to pay for his garden and walkway (in dollars)?

Solution: The entire area will be a 9 feet by 9 feet square, or a total of 81 square feet, 9 of which will be garden. So there will be 72 square feet of walkway. This will cost Johnny $4 \cdot 9 + 6 \cdot 72 = \boxed{468}$ dollars.

8. Arthur wants to build a moat around his castle. The castle sits on a circular plot of land of radius 50 meters. If the moat is to be 10 meters wide all around, what area will the moat cover (in square meters)? Express your answer in terms of π.

Solution: The entire area covered by land and the moat will be a circle with radius 60 meters, which will have an area of 3600π square meters. The land will have an area of 2500π square meters, so the moat will cover $3600\pi - 2500\pi = \boxed{1100\pi}$ square meters.

9. Mr. Bob needs to paint a wall. A bucket of paint can cover 25 square meters. If Mr. Bob's wall is 15 meters wide and 9 meters tall, what is the minimum number of buckets Mr. Bob will need to paint the wall?

Solution: The area of the wall is $15 \cdot 9 = 135$ square meters. That is more than what five buckets can cover, but less than what six buckets can cover. So our answer is $\boxed{6}$.

10. Tom has a 12 feet by 8 feet by 8 feet box. What is the maximum number of iron blocks with dimensions 2 feet by 2 feet by 4 feet that can fit inside Tom's box?

Solution: We divide the volume of the box by the volume of a block, which will give us the number of blocks that can fit inside the box. So we want $\frac{12 \cdot 8 \cdot 8}{2 \cdot 2 \cdot 4} = \boxed{48}$.

Additional Problems

1. A square and an equilateral triangle share a side. What is the ratio of the area of the triangle to the area of the square?

   Solution: Let the side length be s without loss of generality. Then, since the area of an equilateral triangle is $\frac{s^2\sqrt{3}}{4}$ and the area of a square is s^2, the ratio is equal to $\boxed{\dfrac{\sqrt{3}}{4}}$.

2. Circle ω with center O has radius 6. Consider two points P and Q on ω such that $OP = PQ$. Compute the area of triangle OPQ.

   Solution: Notice that the radius of the circle is 6 so $OP = 6$ and $OQ = 6$. Since $OP = PQ$, $PQ = 6$ and OPQ is equilateral. The area of an equilateral triangle is found by $\frac{s^2\sqrt{3}}{4} = \boxed{9\sqrt{3}}$.

3. A rectangle has its length and width increased by 20%. By how many percent will the area increase?

   Solution: Increasing the length and width by 20% is equivalent to multiplying both dimensions by 1.2. As a result, if x and y are the original length and width, the new length and width are $1.2x$ and $1.2y$, respectively. Thus, the new area is $1.2^2 \cdot xy = 1.44xy$. It follows that the area increased by $\boxed{44}$ percent.

4. A square and a triangle have equal perimeters. The lengths of the three sides of the triangle are 6.2 cm, 8.3 cm and 9.5 cm. Compute the area of the square in square centimeters.

   Solution: The perimeter of the triangle is $6.2+8.3+9.5 = 24$ cm, so the square's perimeter must also be 24 cm. Since each side of the square is the same length s, the perimeter must be $4s = 24$ cm. So, $s = \frac{24}{4} = 6$ cm. The area of a square is equal to the side squared, so $A = s^2 = 6^2 = \boxed{36}$ centimeters squared.

5. What is the area (in square units) of a triangle with coordinates $(1,2), (3,0)$, and $(3,3)$?

   Solution: If the base is the side connecting $(3,0)$ and $(3,3)$, then the length of the base is 3. The altitude goes from $(1,2)$ to the base and is perpendicular to the base. Since the x-coordinates of both $(3,0)$ and $(3,3)$ are the same, the base is vertical. Thus, the altitude is horizontal, so the length of the height is the difference of the x-coordinates, or $3 - 1 = 2$. Therefore, the area of the triangle is $A = \frac{1}{2}bh = \frac{1}{2} \cdot 3 \cdot 2 = \boxed{3}$ units squared.

6. A triangle has two side lengths 5 and 7. If the third side is an integer, what is the sum of all possible lengths of the third side?

   Solution: Let x be the third side. By the triangle inequality, $5 + 7 > x$ and $5 + x > 7$. Hence, $2 < x < 12$ so $x = 3, 4, 5, 6, 7, 8, 9, 10, or 11$ and the sum is $\boxed{63}$.

7. Michelle sees a penny 3 feet from her shoes. If her hand is 4 feet above her shoes, how far, in feet, is her hand from the penny?

   Solution: Draw a right triangle with vertices at her hand and shoes and at the penny. The distance from the hand to the penny is the hypotenuse, so it is equal to $\sqrt{3^2 + 4^2} = \sqrt{9 + 16} = \sqrt{25} = \boxed{5}$ feet.

8. Farmer Bob needs to fence in an area of his farm to grow yummy strawberries. Unfortunately, he is on a short budget and can only buy 50 feet of fence to enclose this strawberry patch. Compute the greatest area in which Farmer Bob can grow his strawberries, such that the entire area is fenced in.

Solution: We are given the perimeter of the shape that Farmer Bob will fence in: 50 feet of fence. When given a specified perimeter, a circle will create the greatest possible area. Thus, $2 \cdot \pi \cdot r = 50$, $r = \frac{25}{\pi}$, and the area $\pi \cdot r^2 = \boxed{\frac{625}{\pi}}$

9. How many different isosceles triangles have integer side lengths and perimeter 19?

Solution: We list out the possible sets of side lengths, making sure that each set satisfies the Triangle Inequality: $\{5, 5, 9\}$, $\{6, 6, 7\}$, $\{7, 7, 5\}$, $\{8, 8, 3\}$, $\{9, 9, 1\}$. Thus, there are $\boxed{5}$ different sets of side lengths that satisfy the conditions.

10. Two concentric circles are drawn such that one circle's radius is 3 times the other. If the area between the two circles is 72π, what is the radius of the larger circle? (Two circles are concentric if they share the same center.)

Solution: Let the radius of the larger circle be R, and the radius of the smaller circle be $\frac{R}{3}$. We find that the area between the two circles is $\pi R^2 - \pi \left(\frac{R}{3}\right)^2 = 72\pi$. Cancelling out the π, we find that $\frac{8}{9}R^2 = 72$, so the radius of the larger circle is $\boxed{9}$.

11. The width of a rectangular prism is 3. Its length and height are both 5. What is its surface area?

Solution: The surface area of a rectangular prism is the total area of six faces, or equivalently, twice the total area of the three distinct faces: the rectangles formed by the width and length, width and height, and length and height. Thus, the surface area of the rectangular prism is $2 \cdot (3 \cdot 5 + 3 \cdot 5 + 5 \cdot 5) = \boxed{110}$.

12. If the lengths of two of the sides of an isosceles triangle are 6 and 12, then what is the perimeter of this triangle?

Solution: We must first find which of the two sides is doubled in the isosceles triangle. We see that the triangle cannot have dimensions 6, 6, 12, due to the Triangle Inequality. Thus, the dimensions are 6, 12, 12, and the perimeter of the triangle is $6 + 12 \cdot 2 = \boxed{30}$.

13. The length and width of a rectangle are doubled. What is the percent increase between the new area and the original area?

Solution: Let the length and width of the original rectangle be l and w, respectively. We find that the original area of the rectangle is lw, and the new area is $(2l) \cdot (2w) = 4lw$. Thus, the percent increase is $\dfrac{4lw - lw}{lw} \cdot 100\% = \boxed{300\%}$.

14. A man walks on a path around his house. Starting from his house, he first walks 30 feet east, then 40 feet north, then 60 feet west, then finally 80 feet south. How far, in feet, is he from his house?

Solution: If the man travels 30 feet east and later 60 feet west, he has really traveled 30 feet west. Similarly, traveling 40 feet north and then 80 feet south is equivalent to 40 feet south. Because west and south are perpendicular directions, we can use the Pythagorean Theorem to compute the desired distance: $\sqrt{30^2 + 40^2} = \boxed{50}$ feet.

15. I am trying to draw a right triangle with a hypotenuse length of 15 and a leg length of 4. How long will the other leg be? Express your answer in simplest radical form.

   Solution: From the Pythagorean Theorem, we see that the desired length is $\sqrt{15^2 - 4^2} = \boxed{\sqrt{209}}$.

16. A rectangle has a length of 19 less than two times the width. The diagonal of the rectangle is 13. What is the length of the rectangle?

   Solution: Let the width of the rectangle be w. Then the length of the rectangle is $2w - 19$, and the Pythagorean Theorem shows that $w^2 + (2w - 19)^2 = 13^2$. This simplifies to $5w^2 - 76w + 192 = 0 \implies (5w - 16)(w - 12) = 0$. If $w = \frac{16}{5}$, the length of the rectangle would negative. Therefore, $w = 12$, so the length is $2 \cdot 12 - 19 = \boxed{5}$.

17. Consider a right triangle ABC with $AB = 3, BC = 4$, and $CA = 5$. Consider points P, Q, and R on AB, BC, and CA, respectively such that $PBQR$ is a square. Compute the side length of that square.

   Solution: Consider the diagram on the coordinate plane with B as the origin. Observe that A would be $(3, 0)$ and C would be $(0, 4)$. It follows that the equation for AC is $y = -\frac{4}{3}x + 4$.

   For $PBQR$ to be a square, R must be equidistant from AB and BC (i.e. the x and y axis respectively). Hence, the point R must be in the form (a, a). But R also lies on line AC, so

   $$a = -\frac{4}{3}a + 4 \implies \frac{7}{3}a = 4$$

   It follows that the side length of the square is $a = \boxed{\dfrac{12}{7}}$.

18. Let $ABCDEFGH$ be a cube of side length 5 such that face $ABCD$ is the top face, E is directly below A, F directly below B, G directly below C, and H directly below D. Let

170

S be a unit cube that also has A as a vertex and one of its faces is completely on face $ABCD$. If P is any vertex on S, compute the maximum value of PG^2.

Solution: The unit cube must rest on top of the cube $ABCD$-$EFGH$, and the vertex P that is the farthest from G is the vertex right above A. You can draw the triangle $\triangle PEG$, which is a right triangle. Since the side length of $ABCDEFGH$ is 5, $EG = \sqrt{EF^2 + FG^2} = \sqrt{5^2 + 5^2} = \sqrt{50} = 5\sqrt{2}$ units, and $PE = 5 + 1 = 6$ units, so $PG = \sqrt{PE^2 + EG^2} = \sqrt{(5\sqrt{2})^2 + 6^2} = \sqrt{86}$ units. Thus, $PG^2 = \boxed{86}$.

19. Right triangle $\triangle ABC$ has $AB = 6$, $BC = 8$, and $CA = 10$. What is the length of the altitude from B to side AC? Express your answer as a common fraction.

Solution: The area of a triangle $A = \frac{1}{2}bh$, so $[ABC] = \frac{1}{2}(6)(8) = \frac{1}{2}(10)h$, where h is the length of the altitude to AC. So, $10h = 48$, and $h = \boxed{\dfrac{24}{5}}$.

20. Two circles of radius 3 do not intersect and are contained inside a square. Compute the minimum area of this square.

Solution: In order to minimize the area of the square, the two circles must be tangent to each other and to the square, such that the centers of the circle lie on the diagonal of the square. Let the length of a side of the square be x. From the diagonal, we have $3\sqrt{2} \cdot 2 + 3 \cdot 2 = \sqrt{2}x$. Thus, $x = 6 + 3\sqrt{2}$, and the area of the square is $x^2 = \boxed{54 + 36\sqrt{2}}$.

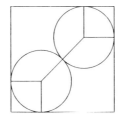

21. A cone sits on the table. It has a height of 30 and a radius of 8. A smaller cone with radius 4 is chopped off from the top (the cut is parallel to the base of the cone). What is the volume of the object that is left over? Express your answer in terms of π.

Solution: Let the volume of the original cone be V. The part of the cone that is chopped off from the top is a cone similar to the original one. The ratio of the dimensions of the chopped cone to the original is $\frac{4}{8} = \frac{1}{2}$. Thus, the ratio of the volume of the chopped cone to the original is $\left(\frac{1}{2}\right)^3 = \frac{1}{8}$. The leftover volume in terms of V is therefore $V - \frac{1}{8}V = \frac{7}{8}V$. We see that $V = \frac{1}{3} \cdot 30 \cdot 8^2\pi = 640\pi$, so the leftover volume is $\frac{7}{8} \cdot 640\pi = \boxed{560\pi}$.

22. Isosceles right triangle ABC has hypotenuse AC. Square $ACDE$ shares an edge of length 1 with triangle ABC. If $ABCDE$ is a convex pentagon, what is the ratio of the perimeter of $ABCDE$ to its area? Express your answer in the form $\frac{a+b\sqrt{c}}{d}$.

Solution: AC must be 1, so $AB = BC = \frac{\sqrt{2}}{2}AC = \frac{\sqrt{2}}{2}$. Since $ACDE$ is a square, $AC = CD = DE = EA = 1$. Consequently, the perimeter of $ABCDE$ is $3 + \sqrt{2}$. The area of $ABCDE$ is equal to the combined areas of ABC and $ACDE$. $[ABC] = \frac{1}{2}AB \cdot BC = \frac{1}{4}$, and $[ACDE] = 1^2 = 1$, so $[ABCDE] = [ABC] + [ACDE] = \frac{5}{4}$. The ratio of the perimeter to the area is

$$\frac{3 + \sqrt{2}}{\frac{5}{4}} = \boxed{\frac{12 + 4\sqrt{2}}{5}}$$

23. Charlie is blowing up balloons for his birth-
day party. Each balloon is a perfect sphere
and has a diameter of 30 cm. After he blows
them up, each of the balloons leaks air at
a rate of 4π cm^3 of air every minute. How
many minutes does it take for the balloons
to shrink to a radius of 12 cm?

Solution: The volume of a sphere is $\frac{4}{3}\pi r^3$, so the initial
volume of the balloons is $\frac{4}{3}\pi 15^3$, and the final volume is
$\frac{4}{3}\pi 12^3$. The total air lost is $\frac{4}{3}\pi(15^3 - 12^3)$. So, since the
rate of air loss is 4π cm^3 per minute, the number of minutes
is $\dfrac{\frac{4}{3}\pi(15^3 - 12^3)}{4\pi} = \dfrac{15^3-12^3}{3} = \boxed{549}$ minutes.

24. A regular hexagon is inscribed in a circle. Another regular
hexagon is circumscribed about the same circle. What is the
ratio of the area of the smaller hexagon to the area of the
larger hexagon?

Solution:

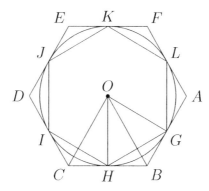

In the above diagram, $ABCDEF$ is the circumscribed hexagon,
and $GHIJKL$ is the inscribed hexagon of circle O. Without
the loss of generality, let the side length of $ABCDEF$ be 1.

Observe that $\triangle OBC$ is an equilateral triangle and OH is the altitude. Since $BC = 1$, $OH = \frac{\sqrt{3}}{2}$.

Next, note that $\triangle OGH$ is also an equilateral triangle. This implies that $GH = \frac{\sqrt{3}}{2}$. The ratio of the areas of two regular hexagons is the square of the ratio of their side lengths, so

$$\frac{[GHIJKL]}{[ABCDEF]} = \left(\frac{\sqrt{3}}{2}\right)^2 = \boxed{\frac{3}{4}}$$

25. Find the area of a triangle with side lengths 17, 25, and 28.

Solution 1: Using Heron's Formula, we have the semiperimeter $s = \frac{1}{2}(17 + 25 + 28) = 35$ and the area of the triangle equal to

$$\sqrt{35(35 - 17)(35 - 25)(35 - 28)} = \sqrt{35 \cdot 18 \cdot 10 \cdot 7}$$

$$= \sqrt{2^2 \cdot 3^2 \cdot 5^2 \cdot 7^2} = 2 \cdot 3 \cdot 5 \cdot 7 = \boxed{210}.$$

Solution 2: Consider two right triangles, one with side lengths 8, 15, and 17, and the other with side lengths 15, 20, and 25. If you put these two right triangles together, with the sides of length 15 coinciding, you get the triangle in this problem. So, the area is equal to the sum of the areas of the two right triangles, which is $\frac{1}{2}(8)(15) + \frac{1}{2}(15)(20) = 60 + 150 = \boxed{210}$.

26. A rhombus is drawn such that one of its angles is $60°$. If the side length of the rhombus is 8, what is the length of the longer diagonal?

Solution: The diagonals of a rhombus are perpendicular to each other and they bisect the angles. As a result, the two diagonals dissect the rhombus into four 30-60-90 triangles. The hypotenuse is equal to the side length of the rhombus, 8. It follows that the length of the longer diagonal is $4\sqrt{3} + 4\sqrt{3} = \boxed{8\sqrt{3}}$.

27. The perpendicular bisectors of sides AB and AC of triangle ABC meet at O. If $\angle COB = 139°$ and $\angle BOA = 97°$, what is $\angle ABC$?

Solution: By definition, O is the circumcenter of triangle ABC. We see that $\angle AOC = 360 - \angle COB - \angle BOA = 124$. So $\angle ABC = \frac{\angle AOC}{2} = \boxed{62}$ degrees.

28. A lamp post and a nearby fire hydrant are 3 meters away from a wall. The lamp post casts a 3.5 meter tall shadow on the wall. The fire hydrant, which is 1 meter tall, casts a 2 meter long shadow along the ground. In meters, how tall is the lamp post?

Solution: The fire hydrant is not tall enough to reach the wall with its shadow, so it casts a 2 meter shadow on the ground. The lamp post, however, is tall enough. If a 1 meter tall fire hydrant leaves a 2 meter shadow on the ground, then it requires 1.5 meters of height to give a 3 meter shadow, which reaches the wall. So the lamp post is $1.5 + 3.5 = \boxed{5}$ meters tall.

29. A spherical balloon i=s being blown up. Its radius is increasing at a rate of $r = \sqrt{12t}$, where t is the time elapsed in seconds and r is the radius in centimeters. What is the volume of the balloon after 3 seconds (in cubic centimeters)? Express your answer in terms of π.

Solution: At $t = 3$ seconds, $r = \sqrt{12 \cdot 3} = 6$. Since the volume of a sphere is $\frac{4}{3}\pi r^3$, we get $\boxed{288\pi}$ as the volume.

30. An equilateral triangle is inscribed in a circle with a radius of 2 inches. In square inches, what is the area of this triangle? Express your answer in simplest radical form.

Solution: If we drop a perpendicular from the center of the circle to one of the sides of the triangle, we create a 30-60-90 triangle with a side length of 2 on the side opposite the right angle. So the side length of the equilateral triangle is $2\sqrt{3}$. Using the formula $A = \frac{\sqrt{3}}{4}s^2$ for the area of an equilateral triangle gives us $A = \boxed{3\sqrt{3}}$.

31. In square units, what is the maximum possible area of a rectangle with a perimeter of 10 units? Express your answer as a common fraction.

   Solution: Trying out rectangles with perimeter 10, such as a 4 by 1 rectangle and a 3 by 2 rectangle, makes it seem as if the closer the sides are in value, the bigger the area. This is true, and a square with side length 2.5 is the maximum case. This has area $\boxed{\dfrac{25}{4}}$.

32. A regular octagon is made by cutting isosceles right triangles out of the corners of a 12 inch by 12 inch square. In inches, what is the perimeter of the octagon? Express your answer in simplest radical form.

   Solution: The hypotenuse of the isosceles right triangles must equal the remaining side of the square after the cut, since we have a regular octagon. So if the length of one side of the octagon is x, we have that $x = 12 - 2(\frac{x}{\sqrt{2}}) \rightarrow x = 12 - x\sqrt{2} \rightarrow x(\sqrt{2}+1) = 12 \rightarrow x = \frac{12}{\sqrt{2}+1}$. Now we rationalize the denominator and get $x = \frac{12(\sqrt{2}-1)}{1}$. The perimeter is simply 8 times this, which is $\boxed{96(\sqrt{2}-1)}$.

33. The volume of a cone and cylinder are equal. If the heights of the two objects are equal, what is the ratio of the radius of the cone to the radius of the cylinder? Express your answer in simplest radical form.

   Solution: The formula for the volume of a cone and cylinder are $\frac{1}{3}\pi r^2 h$ and $\pi r^2 h$, respectively. We know that the heights are equal, so if the radius of our cone is a and the radius of our cylinder is b, we get $\frac{1}{3}a^2 = b^2$. Thus, we have $\frac{a^2}{b^2} = 3 \implies$ $\frac{a}{b} = \boxed{\sqrt{3}}$.

34. A triangle with side lengths 12 units, 16 units, and 20 units is inscribed in a circle. In square units, what is the area of the circle? Express your answer in terms of π.

Solution: The triangle is a right triangle with hypotenuse 20 units. Since the hypotenuse of a right triangle inscribed in a circle is also a diameter of that circle, we get that the radius of the circle is 10 units. So the area is $\boxed{100\pi}$ square units.

Challenge Problems

1. Billy Bob drew a quadrilateral $ABCD$ on the chalkboard. If $\angle BCD + \angle DAB = 180°$, $\angle CDB = 33°$, and $\angle ABC = 61°$, what is the measure of $\angle BCA$?

   Solution: Since $\angle BCD + \angle DAB = 180$ degrees, $ABCD$ is cyclic (it can be inscribed in a circle). Furthermore, $\angle ABC + \angle ADC = 180$ degrees due to the fact that the angles in a quadrilateral add up to 360 degrees. Then

$$\angle ADB + \angle CDB + \angle ABC = 180$$

$$\angle ADB = 180 - 33 - 61 = 86$$

   But due to cyclic quadrilateral $ABCD$, $\angle BCA = \frac{\widehat{AB}}{2} = \angle ADB = \boxed{86}$ degrees.

2. Triangle ABC is an equilateral triangle with a side length of 6. Semicircles with diameters AB, BC, and AC are drawn outside triangle ABC. Circle O circumscribes the new figure. The area of the region outside the 3 semicircles and triangle ABC but inside Circle O can be expressed in the form $\left(a\sqrt{b} - \frac{c}{d}\right)\pi - e\sqrt{f}$, where a, b, c, d, e, f are positive integers, c, and d are relatively prime, and b and f are not divisible by a perfect square of a prime. Compute $a + b + c + d + e + f$.

   Solution: We can compute the desired area by subtracting the area of ABC and the three semicircles from the area of Circle O.

   First, we note that by symmetry, O is also the center of ABC. We then can draw a 30-60-90 triangle ($\triangle ADO$). Since $AD = 3$, we have $OD = \frac{3}{\sqrt{3}} = \sqrt{3}$. It follows that the radius of Circle O is $3 + \sqrt{3}$, so the area of Circle O is $(3 + \sqrt{3})^2\pi = (12 + 6\sqrt{3})\pi$.

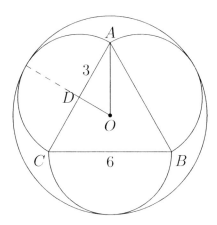

Using 30-60-90 triangles, we see that the height of ABC is $6 \cdot \frac{\sqrt{3}}{2} = 3\sqrt{3}$. As a result, the area of ABC is $\frac{6 \cdot 3\sqrt{3}}{2} = 9\sqrt{3}$.

The combined area of the 3 semicircles is $3 \times \frac{3^2 \pi}{2} = \frac{27\pi}{2}$. Therefore, the area we want to find is

$$(12 + 6\sqrt{3})\pi - \frac{27\pi}{2} - 9\sqrt{3}$$

$$= \left(6\sqrt{3} - \frac{3}{2}\right)\pi - 9\sqrt{3}$$

The answer is then $6 + 3 + 3 + 2 + 9 + 3 = \boxed{26}$.

3. Let D be the point on segment BC of triangle ABC such that $\angle BAD = \angle CAD$. If $AB = 13$, $AC = 14$, and $BD = \frac{65}{9}$, compute the value of the inradius of $\triangle ABC$.

Solution: The given definition of D tells us that AD is the angle bisector of $\angle BAC$. We also have a lot of lengths to work with, so we can use the Angle Bisector Theorem to find CD:

$$\frac{AB}{BD} = \frac{AC}{CD} \implies \frac{13}{\frac{65}{9}} = \frac{14}{CD}$$

$$\implies \frac{9}{5} = \frac{14}{CD}$$

Thus, $CD = \frac{70}{9}$, meaning that $BC = BD + CD = \frac{65}{9} + \frac{70}{9} = \frac{135}{9} = 15$.

We now have all three sides of the triangle, and we want to find the inradius. We know that the inradius times the semiperimeter equals the area of the triangle. And since we have all three sides, we can use Heron's Formula to compute the area. We have $s = \frac{13+14+15}{2} = 21$, so

$$\text{Area of ABC} = \sqrt{21(21-13)(21-14)(21-15)}$$

$$= \sqrt{21(8)(7)(6)} = 84$$

As a result, $21r = 84$, so $r = \boxed{4}$.

4. Anthony and Bob are walking along the figure below, a regular hexagon. They both start out at A and walk at the same speed. Anthony takes the path ABCDEFA, walking around the hexagon once. Bob takes the path ACDEFA, taking a shortcut from A to C. If Anthony takes 6 minutes to walk around the hexagon, how many minutes does it take Bob to walk his path? Express your answer in the form $a + \sqrt{b}$, where a and b are positive integers.

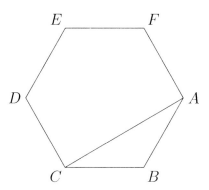

Solution: Draw the perpendicular from B to AC, and have the foot of it be P:

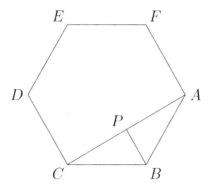

Then, since $\triangle ABC$ is isosceles, BP bisects $\angle ABC$, so $\angle ABP = \angle PBC = 60°$. $\triangle ABP$ and $\triangle PBC$ are both 30-60-90 triangles. Thus, if the side length of the hexagon is 1, then $BP = \frac{1}{2}$ and $AP = CP = \frac{\sqrt{3}}{2}$. So, $AC = \sqrt{3}$, and Bob's path has length $4 + \sqrt{3}$ units. Since Anthony takes 6 minutes to walk his 6 unit path, his speed is 1 unit per minute. Bob travels at the same speed, so it takes him $\boxed{4 + \sqrt{3}}$ minutes to walk his path.

5. A unit cube has a right cylinder cut out of it with height $\frac{7}{13}$ and radius $\frac{1}{3}$. What is the surface area of the remaining object? Express your answer in the form $a + \frac{b}{c}\pi$.

Solution: Let A_{lat} be the lateral surface area of the cylindrical hole, A_{top} be the missing area on the face of the cube out of which the hole was cut, and A_{bottom} be the surface area on the bottom of the hole. Then the total surface area of the cube with the hole $A = A_{cube} + A_{lat} - A_{top} + A_{bottom}$. Since the hole is a right cylinder, $A_{top} = A_{bottom} \implies A = A_{cube} + A_{lat}$. $A_{cube} = 6$ units squared, and $A_{lat} = 2\pi rh = 2\pi(\frac{1}{3})(\frac{7}{13}) = \frac{14}{39}\pi$ units squared, so the total area is $A = \boxed{6 + \frac{14}{39}\pi}$.

Counting and Probability

Warm-up Problems

1. There are 61 families of turtles in Turtleville. Each family has 11 turtles, and each turtle has 3 shells. How many shells are there in Turtleville?

   Solution: We can first count how many turtles there are altogether. With 61 families and 11 turtles, there are $61 \times 11 = 671$ turtles. Since each one has 3 shells there are $671 \times 3 = \boxed{2013}$ shells in Turtleville.

2. A cowboy plans to build a fence to enclose a square pasture. The perimeter of the plot is 96 feet, and he sets posts along the perimeter of the square, keeping adjacent posts 6 feet apart. How many posts will he use to fence the entire plot?

   Solution: If posts are 6 feet apart, there are $\frac{96}{6} = 16$ intervals around the square with length 6. Each interval will contain one post at one of the endpoints, so there are $\boxed{16}$ posts total.

3. Abraham and Belinda are picking out sandwiches for lunch. They must choose a protein (egg, ham, or turkey), a vegetable (lettuce, pickles, or tomatoes), and a type of bread (whole grain, sourdough bread, or white bread). Belinda can only eat egg as a protein. Abraham does not eat tomatoes. However, they must order the same sandwich. How many different sandwiches can they order?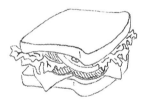

   Solution: Note that both of them must choose egg as the protein, and both cannot choose tomatoes as a vegetable. Thus, there are 2 choices for the vegetable and 3 choices for the bread. This leads to $1 \times 2 \times 3 = \boxed{6}$ different sandwiches.

182

4. If the probability that you get a problem right on any test is 50%, and you take a test with 200 problems, how many questions on the test would you expect to get right?

Solution: If the probability of getting a problem right is 50%, then you can expect to get 50% of the questions on the test right, which is $\frac{50}{100} = \frac{1}{2}$ of the problems, or $\frac{1}{2}(200) = \boxed{100}$ problems.

5. Leonard and Sheldon went to lunch at the Chocolatecake Factory, and they each order 1 item from the menu. The menu has 10 items, and Sheldon refuses to eat what Leonard eats. In how many ways can they have a lunch that satisfies their requirements?

Solution: Leonard has 10 choices to choose from. Sheldon won't choose the same option as Leonard, so he only has 9 choices. Together, they have $10 \cdot 9 = \boxed{90}$ ways to order a satisfying lunch.

6. Batch rolls a regular 6-sided dice and flips 2 fair coins. What is the probability that he rolls a 3 on the die and flips two heads on the coins? Express your answer as a common fraction.

Solution: The probability of Batch rolling a 3 is $\frac{1}{6}$. The probability that he gets heads twice on both flips is $\frac{1}{2} \cdot \frac{1}{2} = \frac{1}{4}$. It follows that the probability that both of these events happen is $\frac{1}{6} \cdot \frac{1}{4} = \boxed{\frac{1}{24}}$.

7. Canton Math Village is made up of a system that contains 3 leaders. Each leader trains 5 generals. Each general trains 1 apprentice. No general is under more than one leader, and no apprentice is taught by more than one general. If everyone in the village is in this system, how many apprentices are there in Canton Math Village?

Solution: There is 1 apprentice per general, and 5 generals per leader. Since we have 3 leaders, we get $3 \cdot 5 \cdot 1 = \boxed{15}$ apprentices.

8. In how many ways can 4 be written as the sum of two or more (not necessarily distinct) positive integers? (Note that order does not matter, so $3 + 1$ is the same as $1 + 3$.)

   Solution: The only possible ways to get 4 using 2 positive integers are $1 + 3$ and $2 + 2$. Using 3 positive integers gets us $1 + 1 + 2$. Finally, we can use 4 with $1 + 1 + 1 + 1$. Thus, our answer is $\boxed{4}$.

9. Rancho San Joaquin Middle School has 1200 students. Each student takes 5 classes a day. Each teacher teaches 4 classes. Each class has 30 students and 1 teacher. How many teachers are there at Rancho San Joaquin Middle School?

   Solution: If there are 1200 students, each student takes 5 classes, and each class has 30 students, then there are $\frac{1200 \cdot 5}{30} = 200$ classes. So, since each teacher teaches 4 classes, there must be $\frac{200}{4} = \boxed{50}$ teachers at Rancho San Joaquin.

10. There are 8 Martian letters in the Martian alphabet. A word in the Martian alphabet consists of 4 Martian letters. How many possible Martian words are there?

    Solution: There are 8 choices for the first letter, 8 choices for the second, 8 for the third, and 8 for the fourth, so in total there are $8 \cdot 8 \cdot 8 \cdot 8 = 8^4 = \boxed{4096}$ possible Martian words.

Additional Problems

1. Compute the number of ways to rearrange the word APPLE.

   Solution: We have 5 letters but 2 are repeated, giving us $\frac{5!}{2!} = \boxed{60}$.

2. How many divisors of 60 are there? A positive integer d is a divisor of 60 if $\frac{60}{d}$ is a positive integer. In particular, 1 and 60 are divisors of 60.

   Solution: The prime factorization of 60 is $2^2 \times 3 \times 5$. A divisor contains any prime to any power less than or equal to the power found in 60. Hence we have $3 \times 2 \times 2 = \boxed{12}$ divisors.

3. Assume every 7-digit whole number is a possible telephone number except those that begin with 0 or 1. What fraction of telephone numbers begin with 9 and end with 0? Express your answer as a common fraction.

   Solution: Since the first digit can't be 0 or 1 in any phone number, there are only 8 choices for the first digit, though there are 10 choices for each of the other 6 digits. So, there are a total of $8 \cdot 10 \cdot 10 \cdot 10 \cdot 10 \cdot 10 \cdot 10 = 8000000$ phone numbers. The restrictions mean that there is only 1 choice for both the first and last digit, though there are still 10 choices for the middle 5 numbers, resulting in a total of $1 \cdot 10 \cdot 10 \cdot 10 \cdot 10 \cdot 10 \cdot 1 = 100000$ phone numbers that have 9 as the first digit and 0 as the last digit. So, the probability is $\frac{100000}{8000000} = \boxed{\frac{1}{80}}$.

4. How many whole numbers between 100 and 400 contain the digit 2?

   Solution: All 100 numbers between 200 and 299 contain the digit 2 in the hundreds digit. For numbers with 1 or 3 as the hundreds digit, we only need to look at the tens and units

185

digit. The only 2-digit combinations that have at least one 2 are 02, 12, 20, 21, 22, 23, 24, 25, 26, 27, 28, 29, 32, 42, 52, 62, 72, 82, or 92, giving us 19 2-digit combinations that have at least one 2. Since there are 2 choices for the hundreds digit, 1 or 3, there are $19 \cdot 2 = 38$ numbers with at least one 2 that aren't between 200 and 299. So, there are a total of $100 + 38 = \boxed{138}$ numbers between 100 and 400.

5. How many ways are there to pick 2 donuts from a box of 12 distinct donuts?

Solution: There are $\binom{12}{2} = \frac{12!}{10! \times 2!} = \frac{12 \times 11}{2} = \boxed{66}$ ways to choose 2 donuts out of 12.

6. Victor is playing a card game, and he is in trouble. In this round, he needs to randomly choose a card from a full deck of 52 cards. He will survive this round if and only if he chooses the queen of spades, or a face card from the suit of hearts. Compute the probability that Victor will not survive this round.

Solution: The cards that Victor can pick to survive are the queen of spades, and the 3 face cards from the suit of hearts: jack, queen, and king. Thus, the probability that he will not survive is $1 - \frac{4}{52} = \boxed{\dfrac{12}{13}}$.

7. If there are 300 cubbyholes and Joe has 301 flyers, what is the probability that one cubby will have at least two flyers?

Solution: Due to the Pigeonhole Principle, recall that after Joe distributes one flyer to each of the 300 cubbyholes, he still has one flyer left. This flyer will have to go into one of the cubbyholes, which all already contain one flyer. Thus, the probability is $\boxed{1}$.

8. John and Eric are going to a ball game. John is always hungry, so they stop by a vendor that sells sandwiches. The

vendor offers a choice of meat from salami, ham, turkey, or chicken; a choice of cheese from American, mozzarella, and Swiss; and a choice of sauce from barbeque, sour cream, mayonnaise, mustard, and dijon. Each sandwich has a meat, a cheese, and a sauce. Unfortunately, John is a picky eater, so he will not eat sandwiches that have both salami and Swiss, chicken and mustard, or mozzarella and dijon. Compute the number of different sandwich combinations John can order.

Solution: The total number of different sandwiches John could have if he were not a picky eater would be $4 \cdot 3 \cdot 5 = 60$ different sandwiches. The different sandwich combinations that John will not eat are salami and Swiss with 5 choices of sauce, chicken and mustard with 3 choices of cheese, and mozzarella and dijon with 4 choices of meat. Thus, the number of different combinations John can order is $60 - (5+3+4) = \boxed{48}$.

9. Ollie the Octopus needs to find socks in the morning. He has bins full of 5 different colors of socks, with each bin having a different color. What is the least number of socks does he have to pull out in the dark so that he can wear eight socks that are all the same color?

Solution: By Pigeonhole Principle, the least number of socks he needs to pull out to guarantee a matching set of socks is $7 \cdot 5 + 1 = \boxed{36}$, or pulling out 7 socks of each color and then one more sock. This last sock is guaranteed to match one of the 5 colors that already have 7 socks.

10. In Mrs. Smith's class, 19 students play piano, 12 students play violin, and 10 students sing. 5 students play piano and sing, 6 students play both piano and violin, 2 students play violin and sing, and 3 students don't do any music. If her class has 33 students, how many students do all three?

Solution: We draw the Venn diagram, and let the number of students who do all three equal x. We know that in the three circles representing piano, violin, and singing, there will be a union of $33 - 3 = 30$ students. After splitting the students into their respective categories, we add the students up in terms of x and set the equation equal to 30. Thus, we find $x = \boxed{2}$ students.

11. 20 couples go to a fancy party. Each person shakes hands with everyone else at the party except for their spouse. Calculate the number of handshakes at the party.

    Solution: Each of the people, of a total of $20 \cdot 2 = 40$ people, shakes hands with $40 - 2 = 38$ people, since they cannot shake hands with themselves or their spouse. Thus, there will be a total of $\frac{40 \cdot 38}{2} = \boxed{760}$ handshakes.

12. A family of 5 members (the mother, father, and three children) are driving to Disneyland in a 5-seat sedan. Only the mother and father can drive, and Bob (one of the three children) becomes carsick when sitting in the passenger's seat. How many comfortable seating arrangements are there?

    Solution: We see that only 2 members of the family can sit in the driver's seat of the sedan. Then, only 3 members can sit in the passenger's seat. The remaining 3 members of the family are ordered in the three back seats. This gives a total of $2 \cdot 3 \cdot 3! = \boxed{36}$ arrangements.

13. OCMC High School has two classes, Calculus and Shakespearean Literature. 125 students are enrolled in Calculus while 18 students are enrolled in Shakespearean Literature. If 8 students are enrolled in both, and obviously a student was take at least one of these classes, how many students attend OCMC High School?

    Solution: Using the Principle of Inclusion and Exclusion, the total number of students $= 125 + 18 - 8 = \boxed{135}$.

14. Kevin fails his biology tests 45% of the time. If Kevin takes 460 tests, on average how many tests will Kevin fail?

    Solution: Kevin fails 45% of the 460 tests, so he fails $\frac{45}{100} \cdot 460 = \boxed{207}$ of the tests.

15. Compute the number of diagonals in a dodecagon.

   Solution: Recall that a dodecagon consists of 12 sides, and thus 12 vertices. The number of lines connecting any two vertices is $\binom{12}{2}$. To compute the number of diagonals, we subtract the number of sides, to get $\binom{12}{2} - 12 = \boxed{54}$ diagonals.

16. Farmer Robert has a rectangular orchard that he wishes to fence, of dimensions 400 feet by 500 feet. He plans to buy high-security fence posts, which are sold individually at his favorite store. To ensure that his orchard is securely fenced, he wants to space his fence posts evenly, one at each corner, every 10 feet. Compute the number of fence posts that he must buy.

   Solution: We compute the number of fence posts he wants for each side of the orchard. By dividing the length of the side by the spacing between the posts, we secure the number of posts of one side, including one of the corner posts. Thus, there will be no overlapping of posts at the corners by adding up the sides. We find that the number of posts Farmer Robert needs to purchase is $2 \cdot (400 \div 10) + 2 \cdot (500 \div 10) = \boxed{180}$.

17. Two dice are thrown. What is the probability that the two-digit number formed will be divisible by 9?

   Solution: Recall that for a number to be divisible by 9, the sum of its digits must be divisible by 9. Thus, we need only find the number of combinations of two numbers such that the sum of the two numbers is a multiple of 9. Since the maximum value of this sum is $6 + 6 = 12$, the sum of the two numbers must be 9. We find that the only possible combinations are: $(3, 6), (4, 5), (5, 4), (6, 3)$. Thus, the probability requested is $\dfrac{4}{6^2} = \boxed{\dfrac{1}{9}}$.

18. Olivia likes to play basketball every other day of the week. If she starts playing on a Tuesday, how many times will she play basketball in two weeks? (The two weeks begin on Tuesday and end on Monday 13 days later.)

Solution: Assume WLOG that Olivia begins playing on a Tuesday that is the first of some month. Then, the end of the two weeks will be on Monday, the 14th of that month. We find the number of odd integers between 1 and 14, inclusive. Thus, Olivia will play basketball $\left\lfloor \dfrac{14}{2} \right\rfloor = \boxed{7}$ days in two weeks.

19. An 8-sided die is rolled (sides numbered 1-8), and then a 6-sided die is rolled. What is the probability that at least 1 prime number is rolled? Express your answer as a common fraction.

Solution: We use complementary counting to find the probability that no prime numbers are rolled. On the 8-sided die, we need to avoid 2, 3, 5, and 7, producing the desired probability $\frac{8-4}{8} = \frac{1}{2}$. On the 6-sided die, we cannot roll 2, 3, and 5, leading to a probability of $\frac{1}{2}$ as well. Thus, the probability that at least one prime number is rolled becomes $1 - \frac{1}{2} \cdot \frac{1}{2} = \boxed{\dfrac{3}{4}}$.

20. Jason and his girlfriend Sally are going on a romantic picnic. Since Jason is very forgetful, he does not know what kinds of sandwiches Sally enjoys. To accommodate for this, he packs 3 types of breads, 5 types of cheeses (including some stinky ones!), and 6 types of meat. However, Sally refuses to have blue cheese with wheat bread, or white bread with roasted ham. How many choices of sandwiches does Sally have, if she must choose one bread, one cheese, and one meat?

Solution: If Sally were not picky, she would have $3 \cdot 5 \cdot 6 = 90$ choices of sandwiches. We see that there are 6 combinations of sandwiches with blue cheese and wheat bread, and

190

5 combinations of sandwiches with white bread and roasted ham. Since these combinations do not overlap, Sally has $90 - (6 + 5) = \boxed{79}$ choices.

21. Emily and Jane are playing rock-paper-scissors. If wins are worth 1 point, ties are worth half a point, and losses are worth 0 points, in how many ways can Jane get 3 points in the first 4 rounds?

    Solution: We represent a win as a W, a tie with a T, and losses with L. Using 4 letters, Jane must get 3 points. She can do so with permutations of WWWL and WWTT. This produces $\dfrac{4!}{1! \cdot 3!} + \dfrac{4!}{2! \cdot 2!} = \boxed{10}$ ways.

22. Seventeen children are in a class, with 10 boys and 7 girls. If all the girls shake each other's hand once, and all the boys shake each other's hands once, how many handshakes take place?

    Solution: We see that there are no overlaps between the girls' and the boys' handshakes, so this becomes simply two separate groups shaking hands within the groups. Thus, there will be $\dbinom{10}{2} + \dbinom{7}{2} = \boxed{66}$ handshakes.

23. Joey is buying donuts. The store only has 2 boxes each of sprinkled donuts, chocolate donuts, and glazed donuts. He wants to buy 3 boxes. In how many ways can he choose the boxes? (Boxes containing the same type of flavor are indistinguishable and the order in which Joey chooses the boxes does not matter.)

    Solution: We see that there are two ways to buy donuts: one box of each flavor, or two boxes of one flavor and one box of another. The first case has 1 way. For the second case, the flavor with two boxes has 3 choices, while the single box has 2 choices. Therefore, Joey has a total number of $1 + 3 \cdot 2 = \boxed{7}$ choices.

24. Larry the zookeeper needs to feed all the animals. There are monkeys, giraffes, elephants, lions, zebras, and hippos. If he must feed the lions either first or last, in how many different orders can he feed the animals?

Solution: Other than the lions, there are 5 more types of animals that Larry has to feed. Ordering these types, there are $5! = 120$ arrangements. Larry must feed the lions either before all 5 types of animals or after all 5 types of animals, so there are 2 ways to insert the lions. Thus, Larry can feed the animals in $120 \cdot 2 = \boxed{240}$ orders.

25. Compute the number of different ten-digit integers that have 4 zeroes, 2 ones, and 4 twos.

Solution: Since we wish to compute ten-digit integers, the first digit in the integer cannot be zero. We look at two different cases. If the first digit is one, we will need to order 4 zeroes, 1 one, and 4 twos to find the other 9 digits of the integer, for a total of $\frac{9!}{4! \cdot 1! \cdot 4!}$ integers. If the first digit is two, we order 4 zeroes, 2 ones, and 3 twos, with $\frac{9!}{4! \cdot 2! \cdot 3!}$ integers. Thus, we have a total of $\frac{9!}{4! \cdot 1! \cdot 4!} + \frac{9!}{4! \cdot 2! \cdot 3!} = 630 + 1260 = \boxed{1890}$ integers.

26. Nancy, Sarah, Janice, Frank, Joseph, and Clark are going to attend a lecture by a prestigious professor, and they have saved six consecutive seats in a row. Unfortunately, Nancy and Clark recently had an argument, so they refuse to sit next to each other. Compute the number of possible seating arrangements.

Solution: We use complementary counting. Recall that without any restrictions on the seating, there would have been a total of $6! = 720$ seating arrangements. We count the number of arrangements where Nancy and Clark are seated next to each other. Since they must be seated next to each other, this is equivalent to ordering five people in a line without any restrictions, or $5! = 120$. However, Nancy and Clark can switch places, thus producing a total of $120 \cdot 2 = 240$

arrangements in which the two are sitting together. Thus, the number of arrangements such that Nancy is not sitting with Clark is $720 - 240 = \boxed{480}$.

27. A spinner has 24 sectors, numbered $1, 2, 3, \cdots 24$. What is the probability that, after spinning the spinner once, it lands on either a prime number or an odd number? Express your answer as a common fraction.

Solution: We compute the number of odd numbers amongst the 24 sectors: $\left\lfloor \dfrac{24}{2} \right\rfloor = 12$ odd integers from 1 to 24, inclusive. All prime numbers from 1 to 24 are included in this set of odd integers except for 2. Thus, the required probability is $\dfrac{12 + 1}{24} = \boxed{\dfrac{13}{24}}$.

28. How many possible ways are there to assign true/false values to Statements 1, 2, and 3 such that the system is consistent?

1. Statement 2 is true
2. Exactly 2 statements are true
3. Statement 1 is false

Solution: If Statement 1 is true, Statement 2 is true, and Statement 3 is false. This system is consistent. Next, if Statement 1 is false, Statement 2 is false and Statement 3 is true. This is also consistent. Hence, there are $\boxed{2}$ ways.

29. William, Kevin, Lohit, and Francis were discussing their possible grades in mathematics class during the grading period. William said, "If I get an A, then Kevin will get an A." Kevin said, "If I get an A, then Lohit will get an A." Lohit said, "If I get an A, then Francis will get an A." All of these statements were true, but only two of the students received an A. Which two received A's?

Solution: Since all the statements are true, then if William gets an A, then Kevin will get an A, so Lohit will get an A,

193

so Francis will get an A, so 4 people will get A's. There-fore, William didn't get an A. If Kevin gets an A, then Lo-hit will get an A, so Francis will get an A, so 3 people will get A's. Thus, Kevin doesn't get an A. If Lohit gets an A, then Francis will get an A, so 2 people get A's. Hence, only ⟨Lohit and Francis⟩ receive A's.

30. There are 30 people in the OCMC Problem Writing Commit-tee. They wrote some algebra problems, geometry problems, and counting problems. At a meeting, 15 people wrote al-gebra problems, 15 people wrote geometry problems, and 20 wrote counting problems. One person wrote only counting problems, 2 wrote only geometry, and 3 wrote only algebra. If 8 people wrote both geometry and counting but not algebra, how many people slacked off and didn't write any problems?

Solution: Consider a Venn Diagram with circles represent-ing the number of people who wrote algebra problems, geom-etry problems, and counting problems. Note that we can fill in some of the spots as shown:

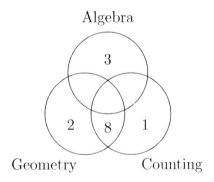

Next, we can count the total number of people in the union of the circles. The entire algebra circle contains 15 people, and the other three sections add up to $2 + 8 + 1 = 11$. Hence, there are $15 + 11 = 26$ people in all. Since 30 people are in the committee, it follows that $\boxed{4}$ people didn't write any problems.

31. How many ways are there for 7 people to sit in a row if there are two couples who want to sit with their spouses?

Solution: Since two couples want to sit with their spouses, each couple will always be sitting with each other. Thus, for each couple, we can substitute one person and take away one chair. We are then left with 5 different people to arrange in a row. This results in $5! = 120$ arrangements. But within each couple, the man and the woman can switch places, so there are ultimately $120 \cdot 2^2 = \boxed{480}$ possible arrangements.

32. Fred the fly is randomly drawing socks out of his drawer without looking. He has 42 red socks, 42 blue socks, 42 green socks, and 13 black socks in his extremely large drawer. How many socks does he need to take out before he can be sure that he has six socks of the same color?

Solution: By the Pigeonhole Principle, if Fred pulls out $4 \cdot 5 + 1 = \boxed{21}$ socks, he must have 6 of the same color. If he pulls out 20 socks, he could have 5 of each color, but the 21st sock must bring one of the colors of socks up to 6.

33. How many ways can you arrange the letters of "Lollipop" if the capital L is different than the 2 lowercase l's?

Solution: Since there are three pairs of identical letters—two o's, two l's, and two p's—and eight letters total, there are $\frac{8!}{2! \cdot 2! \cdot 2!} = \frac{8!}{8} = 7! = \boxed{5040}$ ways to arrange the letters.

34. You are playing a game in which you roll a fair six-sided die each round. If you roll a 2, 3, or 5, you get \$2, \$3, or \$5, respectively. If you roll a 4, you lose \$4. If you roll a 1 or a 6, you don't lose or gain any money. What is the expected value of playing the game once?

Solution: You have a one-sixth chance of rolling each of the numbers. So, you have a one-sixth chance each of gaining \$2, \$3, or \$5 and a one-sixth chance of losing \$4. You also have a one-third chance of neither gaining nor losing anything.

Therefore, the expected value is $\left(\frac{1}{6}\right) \cdot (\$2) + \left(\frac{1}{6}\right) \cdot (\$3) + \left(\frac{1}{6}\right) \cdot (\$5) + \left(\frac{1}{6}\right) \cdot (-\$4) + \left(\frac{1}{3}\right) \cdot (\$0) = \boxed{\$1}$.

35. How many ways are there to seat 3 girls and 6 boys in a row of 9 chairs if the boys must sit in groups of 3? (There must be girls in between the groups of 3; the boys can't sit in one group of 6.)

    Solution: By gender, the girls and boys can sit in one of these arrangements: BBBGBBBGG, BBBGGBBBG, BBBGGG-BBB, GBBBGBBBG, GBBBGGBBB, or GGBBBGBBB. Once the gender order is determined, there are 6! ways to arrange the boys and 3! ways to arrange the girls. So, in total, there are $6 \cdot 6! \cdot 3! = \boxed{25920}$ ways to seat the 9 people.

36. At a restaurant, for a cheeseburger, you can order either white or wheat bread for the buns; only one type of meat; either cheddar, American, or Swiss cheese; and some, none, or all of 5 toppings: lettuce, tomatoes, onions, pickles, and ketchup. How many different cheeseburgers can you order?

    Solution: You have 2 choices for buns, only 1 choice for meat, 3 choices for cheese, and 2 choices for each of the toppings: yes or no. Thus, in total, there are $2 \cdot 1 \cdot 3 \cdot 2^5 = \boxed{192}$ cheeseburgers.

37. Jason is randomly grabbing cookies from a jar that contains chocolate chip cookies, peanut butter cookies, and snicker-doodles. What is the least number of cookies he must grab to ensure he has at least 5 of one kind?

    Solution: By Pigeonhole, if Jason grabs $\boxed{13}$, he must have 5 of one kind. But if he had 12, he could have 4 of every kind.

38. There are 3 oranges, 4 apples, 7 bananas, and 1 watermelon in the pantry. How many ways are there to put the fruits in a line, assuming that fruits of the same kind are indistinguishable?

Solution: This is like arranging a word with letters. The answer is $\frac{(3+4+7+1)!}{3!4!7!1!} = \boxed{900900}$.

39. When an unfair die with faces 1-6 is rolled, the probability of rolling a number X is $\frac{X}{21}$. When this die is rolled twice, what is the probability that the sum of the rolls is even?

Solution: The sum will be even if and only if two even numbers are rolled or two odd numbers are rolled. An even number has probability $\frac{2+4+6}{21} = \frac{4}{7}$ of occurring and an odd number has probability $\frac{1+3+5}{21} = \frac{3}{7}$ of occurring. So the probability that the sum is even is $(\frac{4}{7})^2 + (\frac{3}{7})^2 = \boxed{\frac{25}{49}}$.

40. 4 people are being split into two groups, each group having two people. In how many ways can they be split?

Solution: There are $\binom{4}{2} = 6$ ways to choose the first group, and only 1 way to choose the second group, because you are left with only 2 people. However, the order of the two groups don't matter (choosing AB and then CD is the same as choosing CD and then AB). As a result, we overcounted by a factor of 2, so there are actually only $\boxed{3}$ ways to choose the groups.

41. A 3 inch by 3 inch by 3 inch cube is painted on all sides before being cut up into 27 identical cubes. What fraction of the smaller cubes have less than 2 faces painted? Express your answer as a common fraction.

Solution: The problem asks for the fraction of the smaller cubes that have 1 or 0 faces painted. The only cubes satisfying these conditions are the six in the centers of the six faces and the one in the center of the large cube. So our answer is $\boxed{\frac{7}{27}}$.

42. Robert flips a coin 4 times. What is the probability that he flips heads at least 3 times? Express your answer as a common fraction.

Solution: The probability that he flips exactly 3 heads is $\frac{4}{16}$, as there are 4 ways to pick the placement of the tail. The probability that he flips 4 heads is $\frac{1}{16}$, and adding the two probabilities gives $\boxed{\dfrac{5}{16}}$.

43. Deborah wakes up in the middle of the night to eat a midnight snack. There's a cookie jar next to her bed. Deborah doesn't know which cookie she'll take out since it's dark, but she does know that  the jar contains 6 chocolate chip cookies, 4 oatmeal cookies, and 5 sugar cookies. What is the minimum number of cookies Deborah needs to remove from the jar to ensure that she has two of a single type of cookie?

Solution: Note that it does not matter how many cookies of each kind there are in the jar, but how many kinds. By the Pigeonhole Principle, if Deborah grabs $\boxed{4}$ cookies, she must have two of one type. If she has any less, she doesn't necessarily have two of one type.

44. A fair coin is flipped 5 times. What is the probability that the number of heads that show up will be greater than the number of tails that show up? Express your answer as a common fraction.

Solution: The probability that the number of heads is greater than the number of tails is equal to the probability that the number of tails is greater than the number of heads, by symmetry. Since there is no case where the number of heads and the number of tails are the same since 5 is odd, we get a probability of $\boxed{\dfrac{1}{2}}$.

Challenge Problems

1. How many ways are there to distribute 11 distinct pieces of candy to 3 children?

   Solution: As the children are distinct but the pieces of candy are indistinguishable, we can apply the Stars and Bars method: children are the spaces and pieces of candy are the stars. This gives us an answer $\binom{11+3-1}{3-1} = \binom{13}{2} = \boxed{78}$ ways.

2. Eight people, A, B, C, D, E, F, G and H, finish a race. Given that A beat B, B beat C, C beat D, E beat F, F beat G, and G beat H, how many possible overall rankings are there?

   Solution: These are 2 chains of relations. Let X represent each person in the first chain and Y represent each person in the second. The number of ways to rearrange 4 X's and 4 Y's in a line is $\binom{8}{4} = 70$. Now, the relative order of X's and Y's is already determined by the given rules, so $\boxed{70}$ is the total.

3. Jack and Jill go up the hill with 5 empty pails to a place with 3 rivers. Each of the 5 pails can hold 6 gallons of water. What is the shortest time (in minutes) they need to fill all five pails if each of the rivers gives 1 gallon in 1 minute and no river can give water to more than 1 pail at the same time?

   Solution: Since there are five 6-gallon pails, Jack and Jill must take 30 gallons altogether from the rivers. Each of the three rivers gives 1 gallon per minute, so it would take $\frac{30}{3} = \boxed{10}$ minutes for them to fill up all five pails.

   To see how the rivers can fill up the five pails in only 10 minutes, consider all possible sets of 3 distinct pails. There are 10 such sets, and each pail is part of 6 of the sets. For each set, we fill the pails in the set for exactly one minute.

Then after we finish iterating through the ten sets, we will have spent 10 minutes filling up the pails, and each pail will be full.

4. A group of 10 people are seated in a row of 10 chairs. After leaving their seats for a short break, the people return. In how many ways can the people be seated so that each person is either sitting in her or his previous chair or in a chair adjacent to her or his previous chair?

Solution: Let $P(N)$ represent the number of ways to do this with N people and N chairs. Before the break, let the person sitting in the leftmost chair be A, the person sitting in the second chair from the left be B, and so on, so that the person sitting in the rightmost chair is J. Now, consider the leftmost chair. The person sitting in it after the break must either be A or B. If A sits in the leftmost chair after the break, then the leftmost chair can be ignored, giving $P(9)$ ways to seat the other 9 people. If B sits in the leftmost chair, then A must sit in the second chair from the left, leaving $P(8)$ ways for the other 8 people to sit. So, we have $P(10) = P(9) + P(8)$. This can be applied for any P(N), so we have $P(N) = P(N-1) + P(N-2)$. We can look at the simple cases of $N = 1$ and $N = 2$, giving us $P(1) = 1$ and $P(2) = 2$. So, $P(3) = 3$, $P(4) = 5$, and so on, giving us $P(10) = \boxed{89}$.

5. Victor is buying his baguettes. There are 5 different lengths of baguettes, which are 1, 2, 3, 4, and 5 feet. In how many ways can he make a baguette-chain of length 10 feet with these baguettes?

Solution: We can use careful casework based on the longest baguette Victor buys.

If the longest baguette is 5 feet long, we have $5+5$, $5+4+1$, $5+3+2$, $5+3+1+1$, $5+2+2+1$, $5+2+1+1+1$, or $5+1+1+1+1+1$. There are 7 possibilities in this case.

If the longest baguette is 4 feet long, we have $4+4+2$, $4+4+1+1$, $4+3+3$, $4+3+2+1$, $4+2+2+2$, $4+2+2+1+1$, $4+2+1+1+1+1$, or $4+1+1+1+1+1+1$. There are 8 possibilities in this case.

If the longest baguette is 3 feet long, we have $3+3+3+1$, $3+3+2+2$, $3+3+2+1+1$, $3+3+1+1+1+1$, $3+2+1+1+1+1+1$, or $3+1+1+1+1+1+1+1$. There are 6 possibilities in this case.

If the longest baguette is 2 feet long, there are 5 ways to buy a total of 10 feet. Finally, there is 1 way to buy 10 feet worth of baguettes if we only bought 1 foot baguettes.

Adding up our cases, we get a total of $7+8+6+5+1 = \boxed{27}$ ways.

6. A group of 15 balls are in a line. 5 of these balls are red, and the rest are blue. In how many ways can the line be formed such that a red ball is always next to at least one other red ball?

   Solution: The 5 red balls must be grouped in a group of 5 or in a group of 2 and a group of 3. Consider a row of 10 blue balls. There are 11 'gaps' between the 10 balls (including the ends). Hence, there are 11 ways for the group of 2 red balls to be inserted, and another 11 ways for the group of 3 balls to be inserted. Observe that this takes care of the case in which all 5 balls are together. As a result, the total number of ways to arrange the balls is $11 \cdot 11 = \boxed{121}$.

7. Ana and Banana are playing a game. On a player's turn, they spin a spinner with 5 equal sections that are labeled 1, 2, 3, 4, 5. Ana wins if she spins a 3 or higher, and Banana wins if she spins a prime. The game does not end until one

player wins. What are the chances that Banana wins if she goes second?

Solution: Observe that both players have a $\frac{3}{5}$ chance of winning when it is their turn. Let p be the probability Banana wins if she goes second. Then in the first turn, Ana must spin 2 or lower, and then Banana can win by spinning a prime number. This happens with probability $\frac{2}{5} \cdot \frac{3}{5} = \frac{6}{25}$. Otherwise, Banana does not win when she spins first time, so the probability she wins going second is again p. It follows that

$$p = \frac{6}{25} + \frac{2}{5} \cdot \frac{2}{5} \cdot p$$
$$25p = 6 + 4p$$

Therefore, $p = \boxed{\dfrac{2}{7}}$.

8. A die is rolled until a 5 shows up. What is the expected value of the sum of the rolls?

Solution: Since one roll does not affect the next, we get that the expected value is equal to the amount added when the rolls end now plus the expected value times the probability the rolls do not end plus the average value of the non-ending rolls times the probability the rolls do not end. This becomes $E_x = 5 \cdot \frac{1}{6} + \frac{5}{6} \cdot (E_x + \frac{16}{5})$. Solving, we get $E_x = \boxed{21}$.

9. Al and Ben are playing a card game. Each turn, they pick a random card with replacement from the deck of 52 playing cards. The first person to pick an ace wins. If Al goes first, what is the probability that Ben wins?

Solution: Al has a $\frac{12}{13}$ chance of not picking an ace. Then Ben has a $\frac{1}{13}$ chance of picking an ace. The chances that both do not pick an ace are $\left(\frac{12}{13}\right)^2 = \frac{144}{169}$. So Ben has a $\frac{12}{169} + \frac{12}{169} \cdot \frac{144}{169} + \cdots$ chance of winning, which is simply a geometric sequence that equals $\frac{\frac{12}{169}}{1-\frac{144}{169}} = \boxed{\dfrac{12}{25}}$.

10. A machine tests a person for poison. The machine has a 5% chance of giving a wrong answer. 1000 people use the machine, 10 of whom are poisoned. If the machine tells a person they are poisoned, what is the probability that they really are?

   Solution: We have two cases. Either the person is actually poisoned and the machine is right, or they are not poisoned and the machine is wrong. In the first case, we get probability $\frac{10}{1000} \cdot \frac{95}{100}$, and in the second case, we get $\frac{990}{1000} \cdot \frac{5}{100}$. Dividing the first by the sum of the first and second gives us $\boxed{\dfrac{19}{118}}$.

11. You need to cross a river, but you also need to transport a hungry cat, a hungry mouse, and some cheese. You have one boat, and the boat can only transport you and one other item. If the cat is left with the mouse, the mouse will be devoured. If the mouse is left with the cheese, the cheese will be gone. Compute the minimum number of trips you must make across the river.

   Solution: Realize that the only possible combination that would result in nothing being eaten is having the cat and cheese left together. Thus, on the first trip across the river, you bring the mouse. Then, you go back to the other bank of the river. The third trip, you bring the cat to the other side of the river, and then bring back the mouse. The fifth trip, you take the cheese to the other side, go back to the bank, and bring the mouse. This results in a total of $\boxed{7}$ trips across the river.

Number Sense
Warm-up Problems

1. How many numbers in the set $\{1, 4, 9, 7, -5\}$ cannot be the square of a real number?

   Solution: A square of a real number must be nonnegative, so the only number in the set that isn't the square of a real number is $\boxed{-5}$.

2. What is the product of the integers in the set

   $$\{-15, -22, -12, 16, 0, 17, 195, 57\}?$$

   Solution: There is a zero in that set, so the product is $\boxed{0}$.

3. Compute the sum of the first 5 prime numbers.

   Solution: The first 5 primes are 2, 3, 5, 7, and 11, so the sum is $2 + 3 + 5 + 7 + 11 = \boxed{28}$.

4. What is the greatest integer that divides both 36 and 54?

   Solution: Since $36 = 2^2 \cdot 3^2$ and $54 = 2 \cdot 3^3$, we see that the greatest power of 2 dividing both is 2^1. However, the greatest power of 3 that divides both is 3^2. Thus, the greatest integer that divides into both 36 and 54 is $2 \cdot 3^2 = \boxed{18}$.

5. If the natural number x is divisible by 2, 3, 4, 5, and 6, what is the least possible value of x?

   Solution: This problem asks for the LCM of $2, 3, 4, 5,$ and 6, which is $\boxed{30}$.

6. How many even positive integers divide into 18?

   Solution: There are $\boxed{3}$ such integers: $2, 6, 18$.

204

7. Compute the sum of the factors of 60 that are greater than 12.

Solution: To compute the sum, we can simply list the factors of 60 greater than 12:

$$15, 20, 30, 60$$

The sum of those four factors is $\boxed{125}$.

8. In the month of May there are 31 days. If May 4th is a Sunday, how many Sundays are there in May?

Solution: This problem is equivalent to the number of integers that are 4 (mod 7) from 1 to 31, inclusive. We see that the Sundays are 4, 11, 18, and 25, producing $\boxed{4}$ Sundays in May.

9. What is the sum of all the positive divisors of 24 that are not positive divisors of 18?

Solution: We can simply list the factors of 24, and cross off the ones that are also divisors of 18:

$$\cancel{1}, \cancel{2}, \cancel{3}, \ 4, \cancel{6}, 8, 12, 24$$

Our answer is $4 + 8 + 12 + 24 = \boxed{48}$.

Additional Problems

1. A monkey wants to climb a very tall tree of height 400 feet in the forest to get to the bananas at the top. Every day, it climbs 30 feet, but it slides down 10 feet during the night when it is resting. How many days will the monkey take to secure a banana?

   Solution: At the morning of every day, the monkey finds that it has progressed 20 feet upward from the previous morning. On the last morning, the height of the monkey needs to be at least $400 - 30 = 370$ feet so that it will be able to get to the top of the tree. Thus, we see that it will take $\lceil 370 \div 20 \rceil = 19$ days to reach such a height, and then one more day to reach the top. The monkey will need $\boxed{20}$ days.

2. The digits $2, 3, 4, 7,$ and 9 are used to form the smallest possible five-digit even integer. Compute the value of the integer modulo 100.

   Solution: We wish for the integer to be even, so the units digit must be either 2 or 4. However, we want the integer to be as small as possible, so the first digit must be 2, and the units digit must be 4. Ordering the odd digits from smallest to largest to create the minimum integer, we find that the five-digit integer is 23794, giving the answer $\boxed{94}$.

3. How many even three-digit numbers are divisible by 13?

   Solution: We see that $\lfloor 1000 \div 13 \rfloor = 76$, and that $\lceil 100 \div 13 \rceil = 8$. Thus, we wish to compute the number of even integers from 8 to 76, inclusive. This is $\dfrac{76 - 8}{2} + 1 = \boxed{35}$.

4. If x is the greatest prime factor of 36 and y is the greatest prime factor of 27, what is $x + y$?

   Solution: We find the prime factorizations of $36 = 2^2 \cdot 3^2$, and $27 = 3^3$. Conclude that $x = y = 3$. Thus, $x + y = \boxed{6}$.

5. Annie is playing games of solitaire and listening to music at the same time. Each game of solitaire lasts for 5 minutes. She only wins a game if she hears a Justin Bieber song for some portion of that game. If the Justin Bieber song will play 30 minutes after she begins playing solitaire, Annie will win for the first time when she plays her nth game. What is the value of n?

Solution: Before the Justin Bieber song starts playing, Annie will have played $30 \div 5 = 6$ solitaire games entirely without hearing a Justin Bieber song. The song will begin playing when Annie's 7th game begins; therefore, she will win for the first time when she is on game $\boxed{7}$.

6. Bob has a pile of (at least two) candies that he wants to separate into smaller piles of equal size. When he separates them into piles of 3, he has one candy left over. When he puts them into piles of 5, he has one candy left over. When he puts them into piles of 4, he has one candy left over. What is the minimum number of candies that could have been in the original pile?

Solution: This asks for 1 more than the LCM of 3, 4, and 5, which is $\boxed{61}$.

7. The Fibonacci sequence begins with the terms

$$1, 1, 2, 3, 5, 8, 13, 21, 34, \cdots$$

and each term after the 2nd is defined to be the sum of the previous two terms. What is the remainder when the 1000th term is divided by 5?

Solution: By computing a few terms modulo 5, we see that the remainders repeat every 20 terms, where the last term of the cycle is 0. Thus, since 20 divides 1000, the 1000th term is also $\boxed{0}$ modulo 5.

8. The sum of 3 consecutive nonnegative integers is equal to the sum of 4 consecutive nonnegative integers. Compute the smallest possible value of this sum.

Solution: Let the first sequence be $x, x+1, x+2$ and the second be $y, y+1, y+2, y+3$. Hence, $3x + 3 = 4y + 6$ so $3x - 3 = 4y$. Finally, as the left side is divisible by 3, the right side must be as well, and $y = 3$ and $x = 5$ makes the sum $\boxed{18}$.

9. For integers a and b, let $a \heartsuit b = GCD(a, b) \times LCM(a, b)$, where GCD is the greatest common divisor function and LCM is the least common multiple function. Compute: $6 \heartsuit (14 \heartsuit 32)$.

Solution: We generalize. Let $GCD(a, b) = d$. Then, $a = d \times a'$ and $b = d \times b'$ for some integers a', b'. This makes $GCD(a', b') = 1$ and $LCM(a, b) = LCM(a' \times d, b' \times d) = d \times LCM(a', b')$. The LCM of two relatively prime integers is simply their product. Thus, $GCD(a, b) \times LCM(a, b) = d \times d \times a'b' = a \times b$. This implies that $a \heartsuit b = a \times b$ and we simply calculate $6 \times 14 \times 32 = \boxed{2688}$.

10. If $29x + 11y = 1$ where x and y are integers and $x > 0$, compute the largest possible value of y.

Solution: Taking mod 11, $18x \equiv 1 \bmod 11$. Multiplying both sides by 18's multiplicative inverse, 8, we see that $18 \times 8 \times x \equiv 8 \bmod 11$ so $144x \equiv 8 \bmod 11$ and $x \equiv 8 \bmod 11$. Also, to maximize y, we take the smallest positive value of x: 8. Thus, $x = 8$ and $\boxed{y = -3}$.

11. A terrible analog clock loses 5 minutes per hour. If the correct time right now is coincidentally shown on the clock, after how many hours will the clock display the correct time again?

Solution: Losing 5 minutes per hour means that each hour in real time becomes only 55 minutes in the clock's time. So, after H hours pass, the clock only shows $\frac{55}{60}H = \frac{11}{12}H$ hours passed. For the clock to show the correct time, it must be off by $12N$ hours, where N is an integer. So, $H - \frac{11}{12}H = 12N$, and $\frac{1}{12}H = 12N$, and $H = 144N$, so the smallest amount of hours is $\boxed{144}$.

12. In a certain year, January had exactly four Tuesdays and four Saturdays. On what day of the week did January 1 fall that year?

    Solution: The month of January has 31 days, which is equal to 4 weeks and 3 days. This means that 4 out of the 7 days in the week must occur 4 times, and 3 of them must occur 5 times. The 3 days of the week that occur 5 times are the first three days in the month. Since those 3 days must be consecutive, and none of them can be Tuesday or Saturday, those three days must be Wednesday, Thursday, and Friday. Thus, January 1 must be on $\boxed{\text{Wednesday}}$.

13. For every dress that a seamstress sews, she earns \$8. She is given an extra \$4 for every 7 dresses she sews. How many dresses must she sew to earn \$520?

    Solution: For every 7 dresses she sews, she earns $7(\$8) + (\$4) = \$60$. So, if she sews 7 dresses 8 times, for a total of 56 dresses, she earns $8(\$60) = \$480$. She needs \$40 more to earn \$520, so she needs to sew $\frac{\$40}{\$8} = 5$ dresses more. So, she needs to sew $56 + 5 = \boxed{61}$ dresses in total.

14. Some students are seated at a table when a bag full of $10,000$ candies is passed. If Bob gets both the first piece of candy and the last piece of candy, what is the maximum possible number of students seated around the table?

    Solution: If Bob gets the first and last piece of candy, then 9999 candies is enough to give the same number of candies to each person at the table. So, the number of people N is a factor of 9999. The largest factor of 9999 is 9999, so the maximum number of people at the table is $\boxed{9999}$.

15. What is the sum of the digits of the sum of the digits of the sum of the digits of 2^{18}?

Solution: $2^{18} = 262144$, so the sum of the digits is $2+6+2+1+4+4 = 19$, the sum of the sum of the digits is $1+9 = 10$, and the sum of the sum of the sum of the digits is $1+0 = \boxed{1}$.

16. Compute the number of zeroes at the end of $1^1 \cdot 2^2 \cdot 3^3 \cdot 4^4 \cdots 42^{42} \cdot 43^{43}$.

    Solution: We need to compute the number of zeroes at the end, or equivalently, the number of powers of 5, since there are more than enough powers of 2 in this product. We see that the number of powers of 5 is equivalent to the sum of the multiples of only one power of 5, plus twice the sum of the multiples of exactly two powers of 5, etc. Thus, the number of powers of 5 is $(5 + 10 + 15 + \cdots + 35 + 40) + (25) = 205$, producing $\boxed{205}$ zeroes at the end.

17. What is the smallest positive integer with four factors?

    Solution: An integer with four factors has to be a cubic, or a product of two distinct primes. In the first case, the smallest integer would be $2^3 = 8$, and the minimum integer for the second case is $2 \cdot 3 = 6$. Thus, the smallest positive integer with four factors is $\boxed{6}$.

18. In a total of 15 numbers, the median is 38. The largest number is 93. If the largest number is changed to 97, what is the median?

    Solution: Recall that the median in a set with 15 elements is simply the 8th largest element. Thus, increasing the largest number in the set will allow that number to remain the largest number in the set, not affecting the median, which remains $\boxed{38}$.

19. What is the greatest positive integer that divides both $37^2 + 1$ and $37^3 + 37^2 + 1$?

    Solution: We wish to find the greatest common divisor of $37^2 + 1$ and $37^3 + 37^2 + 1$. By the Euclidean Algorithm, this value is also the greatest common divisor of $37^2 + 1$ and $(37^3 + 37^2 + 1) - (37^2 + 1) = 37^3$. However, $37^2 + 1$ and 37^3 are relatively prime; therefore, the greatest common divisor is $\boxed{1}$.

20. How many integers are factors of 1008, but not of 36?

   Solution: If we factor 1008, we get $2^4 \cdot 3^2 \cdot 7$. So, in total, 1008 has $(4+1)(2+1)(1+1) = 30$ factors. Factoring 36 gives us $2^2 \cdot 3^2$, so 36 has 9 factors. Since 1008 is divisible by 36, all of the factors of 36 are also factors of 1008. This results in $30 - 9 = \boxed{21}$ integers that are factors of 1008, but not of 36.

21. What is the smallest integer greater than 3 that leaves a remainder of 3 when divided by 4, 5, 6, and 7?

   Solution: Let the integer be n. Since n leaves a remainder of 3 when divided by 4, 5, 6, and 7, the number $(n - 3)$ must be divisible by 4, 5, 6, and 7. $n > 3$, so $(n - 3)$ must be a positive integer. To find the smallest possible value of n, we must find the smallest $(n - 3)$. Since $(n - 3)$ must be a multiple of 4, 5, 6, and 7, the smallest $(n - 3)$ is the LCM of 4, 5, 6, and 7, which is 420. This means that the smallest n is $\boxed{423}$.

22. If the digits of a certain two digit integer are reversed, the new number is 36 more than the original integer. The sum of the digits of the original number is 10. What is the new number?

   Solution: Let the original number be $\overline{ab} = 10a + b$, and the new number be $\overline{ba} = 10b + a$, where a and b are digits between 0 and 9 inclusive. So, $(10b + a) - (10a + b) = 36 = 9b - 9a = 9(b - a)$, and $b - a = 4$. Since $a + b = 10$, $b = \frac{1}{2}((a + b) + (b - a)) = \frac{1}{2}(10 + 4) = 7$, and $a = (a + b) - b = 10 - 7 = 3$. We see that the new number is $\overline{ba} = \boxed{73}$.

23. What is the value of 6 times 9 in base 10 when expressed in base 13?

   Solution: 6 times 9 in base 10 is 54. If 54 is $\overline{ab}_{13}$ in base 13, then $54 = 13a + b$, where a and b are nonnegative integers are less than 13. Since $4 \cdot 13 = 52 < 54 < 65 = 5 \cdot 13$, $a = 4$, and $b = 54 - 4 \cdot 3 = 2$. So, 54 in base 13 is $\boxed{42_{13}}$.

24. Harry the snail is climbing over a small hill. Every day, he moves forward 6 inches, and every night, he slides down 2 inches in his sleep. How many days would it take him to climb up one side and down the other side if the hill is 10 feet tall?

   Solution: When going up the hill, he climbs up 6 inches and slides down 2 inches in a day, so he moves a total of 4 inches up the hill every day. Therefore2, it would take him $\frac{120}{4} = 30$ days to climb up the hill. When climbing down the hill, he climbs down 6 inches, and then slides down another 2 inches, resulting in a total of 8 inches per day. So, it takes him $\frac{120}{8} = 15$ days to climb back down. In total, he needs $\boxed{45}$ days to climb over the hill.

25. What is the remainder when the 55th Fibonacci number is divided by 4, assuming that the first Fibonacci number is 1?

   Solution: The Fibonacci numbers are cyclic mod 4 with a period of 6. 55 is 1 mod 6 so our answer is the first number in the cycle which is $\boxed{1}$.

26. Victor is buying baguettes. He wants to buy a number of baguettes that would have 2 left over if split among 3, 1 left over if split among 5, 3 left over if split among 8, and 9 left over if split among 11. What is the smallest number of baguettes he can buy?

Solution: Combining the first two conditions gets us 11 if split among 15. Combining that with the third condition gets us 11 if split among 120, and adding the last condition gets us 251 if split among 1320. So our answer is $\boxed{251}$.

27. When a certain number not divisible by 10 between 5000 and 6000 exclusive is multiplied by 9, its last three digits remain the same. What is the number?

    Solution: We are looking for a number where

    $$9x \equiv x \quad (\text{mod } 1000) \rightarrow 8x \equiv 0 \quad (\text{mod } 1000)$$

    Since $10 \nmid x$, the only possible solution is $x \equiv 125 \pmod{1000} \rightarrow x = \boxed{5125}$.

28. Is 8573624457820152763495 divisible by 9? If not, what is the remainder?

    Solution: Adding up all the digits of that number gives us 103, which is equivalent to 4 (mod 9). So our answer is $\boxed{4}$, and the number is not divisible by 9.

29. If a, b, and c are positive integers and $abc = 12$, how many possible values of ab are there?

    Solution: $ab = \frac{12}{c}$, so ab can be any divisor of 12, of which there are $\boxed{6}$.

30. The mean, median, and mode of the following set are all identical: $17, n, 13, 17, 13, 17, 22$. What is n?

    Solution: The mode must be 17, so we need to set the mean equal to 17. We have $\frac{99+n}{7} = 17 \rightarrow n = \boxed{20}$.

31. Elizabeth has an older brother and older sister. The sum of the three siblings' ages is 8. The product of their ages is 12. How old is Elizabeth?

Solution: Since their ages are all integer values, and Elizabeth's siblings are both older than her, there are only three possibilities that sum to 8: $(2,3,3)$, $(1,2,5)$ and $(1,3,4)$. Only $(1,3,4)$ gives a product of 12, so Elizabeth's age is $\boxed{1}$.

32. The median of a set of five positive integers is 7. The mean of the set is also 7. The mode of the set is 3. What is the greatest number that could be in the set?

    Solution: The set must be something like this: $\{3,3,7,x,y\}$. We want to maximize y, so we set x to be as low as possible, namely 8. If $x = 7$, we wouldn't have a distinct mode. So using $x = 8$ and the fact that the mean is 7, we get that $y = \boxed{14}$.

33. If Jerry takes a one-day break from school after every 2 school days, Robi takes a one-day break after every 3 school days, and school starts on a Monday, what is the first day both Jerry and Robi will be both take a one-day break on the same day? Assume that there are no weekends, holidays, etc.

    Solution: Jerry takes the 3rd day off of every 3 days, and Robi takes the 4th day off of every 4 days. The LCM of 3 and 4 is 12, so on the twelfth day they will both be taking a break. The first day is a Monday, and so the twelfth day is a $\boxed{\text{Friday}}$. (Note: These are very bad students. Do not follow their example.)

Challenge Problems

1. Let N be the product of all positive integers that divide into 600 but not 420. How many terminal zeroes does N have?

   Solution: Since $600 = 2^3 \cdot 3 \cdot 5^2$, 600 has $4 \cdot 2 \cdot 3 = 24$ factors. Thus, the product of all of them is 600^{12}. Next, we must divide out the factors of both 600 and 420. Observe that this is equivalent to dividing out the factors of $\gcd(600, 420) = 60$. Since $60 = 2^2 \cdot 3 \cdot 5$, the product of all factors of 60 is 60^6.

   It follows that

   $$N = \frac{600^{12}}{60^6} = \left(\frac{600^2}{60}\right)^6 = 6000^6$$

   so the number of terminal zeroes is $3 \cdot 6 = \boxed{18}$.

2. Let n be a positive integer less than 40 that has more than 4 factors but less than 9 factors. Compute the number of possible values that n can take on.

   Solution: If n has 5 factors, then n must be of the form p^4, where p is a prime. So, $p = 2$ gives $n = 16$, but $p = 3$ gives $n = 81 > 40$. If n has 6 factors, then n either is of the form p^5 or $p \times q^2$, where p and q are different primes. For $n = p^5$, $p = 2$ gives $n = 2^5 = 32$. For $n = p \times q^2$, $(p, q) = (2, 3)$ gives $n = 2 \times 3^2 = 18$, $(p, q) = (3, 2)$ gives $n = 3 \times 2^2 = 12$, $(p, q) = (5, 2)$ gives $n = 5 \times 2^2 = 20$, and $(p, q) = (7, 2)$ gives $n = 7 \times 2^2 = 28$. If n has 7 factors, then $n = p^6$, but even $p = 2$ gives $n = 2^6 = 64$, which is too big. If n has 8 factors, then n must be of the form p^7, $p \times q^3$, or $p \times q \times r$. $n = p^7$ is too big. For $n = p \times q^3$, $(p, q) = (3, 2)$ gives $n = 3 \times 2^3 = 24$. For $n = p \times q \times r$, $(p, q, r) = (2, 3, 5)$ gives $n = 2 \times 3 \times 5 = 30$. The numbers that work are 12, 16, 18, 20, 24, 28, 30, and 32, so there are $\boxed{8}$ possible values of n.

3. Compute the first prime k such that 1 more than the product of all the primes less than or equal to k is not prime.

Solution: At first, this seems impossible, as trying $k = 2, 3, 5, 7, 11$ all seem to give primes. However, $k = \boxed{13}$ makes the product to be 30030 and the total to be 30031. Notice that $30031 = 59 \times 509$ and is not prime.

4. When the binomial coefficient $\dbinom{125}{32}$ is written out in base 10, how many zeroes are at the rightmost end?

Solution: Recall that

$$\binom{125}{32} = \frac{125!}{93! \cdot 32!} = \frac{125 \cdot 124 \cdot 123 \cdots 95 \cdot 94}{32 \cdot 31 \cdot 30 \cdots 2 \cdot 1}$$

To compute the number of zeroes at the end, we need only find the power of 5 in the evaluated coefficient, since we can find that there are enough powers of 2 for every power of 5. We see that in the numerator there are $\left(\lfloor \frac{125}{5} \rfloor + \lfloor \frac{125}{25} \rfloor + \lfloor \frac{125}{125} \rfloor\right) - \left(\lfloor \frac{93}{5} \rfloor + \lfloor \frac{93}{25} \rfloor + \lfloor \frac{93}{125} \rfloor\right) = 10$ powers of 5. In the denominator, we find $\lfloor \frac{32}{5} \rfloor + \lfloor \frac{32}{25} \rfloor = 7$ powers of 5. Thus, the binomial coefficient has $10 - 7 = 3$ powers of 5, and thus $\boxed{3}$ zeroes at the right end.

5. What is the units digit of $2013^{2014^{2015}}$?

Solution: The units digit of 2013^n cycles through 3, 9, 7, and 1. If n is divisible by 4, the units digit is 1. If n leaves a remainder of 1 when divided by 4, the units digit is 3. If the remainder is 2, the units digit is 9. If the remainder is 3, the units digit is 7. In this case, $n = 2014^{2015}$. Since 2014 is divisible by 2, 2014^{2015} is divisible by 4, so the units digit of $2013^{2014^{2015}}$ is $\boxed{1}$.

6. In the world of Orange, peels are worth 4 cents and tangerines are worth 7 cents. What is the minimum total number of peels and tangerines needed to make 50 cents?

216

Solution: Let p be the number of peels and t be the number of tangerines. We have $4p + 7t = 50$ and we need to minimize $p+t$. We can first try to find the solutions to the Diophantine equation $4p + 7t = 50$. Taking this equation modulo 4, we have $3t \equiv 2 \pmod 4$. Multiplying both sides of the congruence by 3, we get $t \equiv 2 \pmod 4$. Thus, $t = 2$ is a solution, giving us $p = 9$.

We see that in general, the integer solutions, (p, t), are in the form $(9 - 7n, 2 + 4n)$, where n is an integer. We want to minimize $p + t = 9 - 7n + 2 + 4n = 11 - 3n$. We now can try to find bounds for n, knowing that both p and t must be nonnegative integers. Then $9 - 7n \geq 0 \implies n \leq 1$. Similarly, $2 + 4n \geq 0 \implies n \geq 0$. As a result, $0 \leq n \leq 1$, implying that the minimum value of $p + t = 11 - 3n$ occurs at $n = 1$, giving us $\boxed{8}$. (The minimum is achieved with 2 peels and 6 tangerines.)

7. For how many positive integers n ≥ 2 is $n^2 - 3n + 2$ a prime number?

Solution: We can factor $n^2 - 3n + 2$ into $(n - 1)(n - 2)$. Since n ≥ 2, neither $(n - 1)$ nor $(n - 2)$ is negative. So, since we want $(n - 1)(n - 2)$ to be a prime number, either $(n - 1)$ or $(n - 2)$ must be equal to 1. If $(n - 1) = 1$, then $n = 2$. Plugging this in, we get $n^2 - 3n + 2 = (n - 1)(n - 2) = 1 \cdot 0 = 0 \implies n = 2$ doesn't work. If $(n - 2) = 1$, $n = 3$, and $n^2 - 3n + 2 = (n - 1)(n - 2) = 2 \cdot 1 = 2$. Thus, $n = 3$ is the only solution, and only $\boxed{1}$ value of n satisfies the conditions.

8. The sum of the reciprocals of three consecutive integers is equal to $\frac{47}{60}$. What is the smallest of these three integers?

Solution: $\frac{1}{a} + \frac{1}{b} + \frac{1}{c} = \frac{ab+bc+ac}{abc}$. Thus, the denominator, in this case 60, should either be abc or a factor of abc. Since a, b, and c are three consecutive integers, $3 \cdot 4 \cdot 5 = 60$ and $4 \cdot 5 \cdot 6 = 120$ give us two easily checked possibilities. $\frac{1}{3} + \frac{1}{4} + \frac{1}{5} = \frac{12+20+15}{60} = \frac{47}{60}$, so the consecutive integers are 3, 4, and 5, giving us an answer of $\boxed{3}$.

9. How many of the factors of 237,600 are perfect squares?

Solution: We see that the prime factorization of $237600 = 2^5 \cdot 3^3 \cdot 5^2 \cdot 11^1$. If a number N is a factor of 237600 and a perfect square, then $N = 2^{2a} \cdot 3^{2b} \cdot 5^{2c} \cdot 11^{2d}$, where a, b, c, and d are nonnegative integers, and $2a \leq 5$, $2b \leq 3$, $2c \leq 2$, and $2d \leq 1$. Thus, $a \leq 2$, $b \leq 1$, $c \leq 1$, and $d = 0$. So, there are in total $(2+1)(1+1)(1+1)(0+1) = 12$ ways to choose a, b, c, and $d \implies \boxed{12}$ possible values of N.

10. A positive integer has 20 factors. What is the sum of the smallest 2 numbers that fit that condition?

Solution: There are 4 cases for a number to have 20 factors. The number N can either equal a^{19}, ab^9, a^3b^4, or abc^4, where a, b, and c are prime. The smallest number for the first case is 2^{19}, which is much too big to be one of the smallest numbers. The smallest number for the second case is $2^9 \cdot 3 = 1536$, which is still too big. The smallest numbers for the third case are $2^4 \cdot 3^3 = 432$ and $2^3 \cdot 3^4 = 648$, and the smallest numbers for the fourth case are $2^4 \cdot 3 \cdot 5 = 240$ and $2^4 \cdot 3 \cdot 7 = 336$. Thus, the smallest numbers are 240 and 336, and the answer is $\boxed{576}$.

11. How many factors of 19404 are squares?

Solution: We get the prime factorization of 19404 which is $2^2 \cdot 3^2 \cdot 7^2 \cdot 11$ which means all factors of 19404 are in the form $2^a \cdot 3^b \cdot 7^c \cdot 11^d$. For a factor to be a perfect square, all of the exponents must be even. It follows that there are $2 \cdot 2 \cdot 2 \cdot 1 = \boxed{8}$ perfect square factors.

12. Eric is driving around his neighborhood. He looks at his odometer, which displays a 5-digit palindrome less than 20,000. 11 miles later, it displays another palindrome. What is his odometer reading 15 miles after that?

Solution: It seems impossible to get another palindrome in 11 miles, since that usually only changes the last two digits.

But when we use something very close to 20000 like 19991, adding 11 changes all the digits, namely to 20002, which is a palindrome. Adding 15 miles gives us $\boxed{20017}$.

Review

Problems

1. Evaluate: $4^2 - 2 \times 4 \times 3 + 3^2$.

   Solution: $4^2 - 2 \times 4 \times 3 + 3^2 = 16 - 8 \times 3 + 9 = 16 - 24 + 9 = -8 + 9 = \boxed{1}$.

2. A circular disk with radius 1 is placed randomly on a circular table with radius 7 so that the center of the disk is on the table. What is the probability that the whole disk is on the table, with no part hanging off?

   Solution: For the entire disk to be on the table, the center must be at least 1 unit from the edge of the table. Thus, the boundary of where the center of the disk can be is 1 unit inwards from the edge of the table, which is a circle with radius 6:

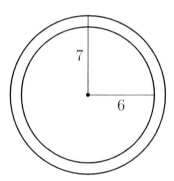

   So, the area of the region where the center of the disk can be is 36π units squared, and the area of the table is 49π units squared, so the probability is $\boxed{\dfrac{36}{49}}$.

3. Victor is eating his baguettes. He takes 3 inch bites or 5 inch bites. How many ways can he get through a 2 foot baguette?

Solution: Victor can either take eight 3 inch bites or three 3 inch bites and three 5 inch bites. There is 1 way to order eight 3 inch bites and $\binom{6}{3}$ ways to order three 3 inch bites and three 5 inch bites. Adding those gives us $\boxed{21}$.

4. If $a = 32$, $b = 21$, and $c = 29$, what is $(a + c)(a - b)(a - c)$?

Solution: $(a+c)(a-b)(a-c) = (32+29)(32-21)(32-29) = (61)(11)(3) = (183)(11) = \boxed{2013}$.

5. If the sum $1 + 2 + \cdots + n = 1540$, what is the value of n?

Solution: There are two ways to get around this problem. Both ways use the fact that $1 + 2 + \cdots + n = \frac{n(n+1)}{2}$. We can either factor 1540 into $4 \cdot 5 \cdot 7 \cdot 11 = 55 \cdot 28 = \frac{55 \cdot 56}{2} \Rightarrow n = 55$. A stricter way to solve this problem would be to use $1+2+\cdots+n = \frac{n(n+1)}{2} \Rightarrow n(n+1) = 3080 \Rightarrow n^2 + n - 3080 = 0 \Rightarrow n = \boxed{55}$

6. How many integers from 1 to 500, including 1 and 500, are multiples of 2 but not multiples of 5?

Solution: We find the number of multiples of 2 and subtract the number of multiples of 10, producing all of the multiples of 2 that are not divisible by 5. Thus, the desired value is $\left\lfloor \frac{500}{2} \right\rfloor - \left\lfloor \frac{500}{10} \right\rfloor = 250 - 50 = \boxed{200}$.

7. A coin has a probability of $\frac{1}{3}$ that it will show heads when flipped, and a probability of $\frac{2}{3}$ that it shows tails. The coin is flipped twice. What is the probability that it showed tails both times? Express your answer as a common fraction.

Solution: The coin has a $\frac{2}{3}$ chance of being tails the first time and $\frac{2}{3}$ chance of being tails the second time. So the probability that these both happen is $\frac{2}{3} \cdot \frac{2}{3} = \boxed{\frac{4}{9}}$.

8. Jane and Suzy are playing basketball. Jane makes her free throws $\frac{2}{3}$ of the time while Suzy only makes hers $\frac{1}{2}$ of the time. If Jane shoots 2 free throws and Suzy shoots 3, what is the probability that they don't make any free throws? Express your answer as a common fraction.

Solution: The probability that they both make no shots is the probability that Jane makes none times the probability that Suzy makes none which is

$$\left(\frac{1}{3}\right)^2 \cdot \left(\frac{1}{2}\right)^3 = \boxed{\frac{1}{72}}$$

9. Triangle ABC has a right angle at B. D is the foot of the altitude from B. If BD is 12 and $\frac{AB}{BC} = \frac{3}{4}$, what is AB?

Solution: Let the length of AB be x. Then we see that BC is $\frac{4}{3}x$, so AC is $\sqrt{x^2 + (\frac{4}{3}x)^2} = \frac{5}{3}x$. Recall that $AB \cdot BC = BD \cdot AC$, so $x \cdot \frac{4}{3}x = 12 \cdot \frac{5}{3}x$, so $x = AB = \boxed{15}$.

10. Suppose Jennifer writes a list of 100 positive integers less than 100. If she writes another list with each number from the first list subtracted from 100, then what is the sum of the numbers on both of her lists?

Solution: Let a be an integer from Jennifer's first list. On the second list, the integer corresponding to this number is $100 - a$. When adding the two lists, the corresponding sum of the two integers is 100. The same is true for each of the other 99 integers in the first list, so the required sum is $100 \cdot 100 = \boxed{10000}$.

11. In Mr. Faraday's classroom, everyone has at least one calculator. The only types of calculators are TI-84's and TI-89's.

Eighteen students have TI-84's, and two-thirds of all the students have TI-89's. If there are 30 students in Mr. Faraday's class, how many students have both a TI-84 and a TI-89?

Solution: Two-thirds of 30 is 20. Since everybody has at least one calculator, we add 20 and 18 to get 38, which is 8 more than 30, which means $\boxed{8}$ students have both calculators.

12. A number is called *sphinxlike* if it has 3 digits and the first and last digits are identical. How many *sphinxlike* numbers are there?

Solution: The first digit cannot be 0, or else it would not be a 3 digit number. So we have 9 choices for the first and last digit and 10 choices for the middle digit, which gives us $\boxed{90}$ possible *sphinxlike* numbers.

13. Ann and Brenda are having a picnic. They want to make enough sandwiches to feed the both of them without having leftovers. Ann needs to eat 3 sandwiches but can stuff herself to eat 5. Brenda needs to eat 4, but can stuff herself to eat 6. What is the difference between the largest and smallest amount of sandwiches they can bring?

Solution: If they both stuff, they can eat 11. If they both eat their minimum, they eat 7. We find that $11 - 7 = \boxed{4}$ sandwiches.

14. John's computer has 2 GB of RAM (1 GB = 1000 MB). His computer requires 509 MB for operating system needs, and he is currently running his internet browser, which takes up 129 MB. He is also running a game that takes up 856 MB. How much RAM is still open?

Solution: We get $2000 - 509 - 129 - 856 = \boxed{506}$.

15. Tom's batting averages for his first three seasons are 0.267, 0.243, and 0.256. The mean of his batting averages for his first four seasons is 0.261. What is his batting average for his fourth season? Express your answer as a decimal rounded to the nearest thousandth.

Solution: If the batting average of his fourth season is x, we get that $\dfrac{0.267 + 0.243 + 0.256 + x}{4} = 0.261$. Multiplying each side of the equation by four gives us $0.766 + x = 1.044 \rightarrow x = \boxed{0.278}$.

16. Aaron teaches tennis lessons and earns 200 dollars a week. However, he has to restring his racquets at the end of every two weeks and this costs him 40 dollars. He wants to buy a 920 dollar laptop. If he starts out with 0 dollars, how many weeks will this take him?

Solution: Aaron will have \$360 by the end of two weeks, and he will have \$720 by the end of four weeks. Aaron works one more week and gets \$200 more, which brings his total amount of money to \$920. So Aaron needs to work $\boxed{5}$ weeks.

17. A sports store sells basketballs and baseballs. Each basketball costs 10 dollars and each baseball costs 6 dollars. Avery buys 13 basketballs and baseballs altogether at the store, and paid 110 dollars total. How many basketballs did he buy?

Solution: Avery bought b basketballs and $13 - b$ baseballs. We have $10b + 6(13 - b) = 110$. Solving the equation we get $b = \boxed{8}$.

18. Every student in the All-Girls Tournament shakes hands with every other student, for a total of 153 handshakes. How many students are in the tournament?

Solution: If there are N students in the tournament, then there are $\binom{N}{2} = \frac{N(N-1)}{2}$ handshakes in total. So, $\frac{N(N-1)}{2} = 153$, and $N(N-1) = 306$. $N = 18$ satisfies this equation, so there are $\boxed{18}$ students in the tournament.

19. Two dice are thrown. What is the probability that the sum of the numbers on the top faces is even? Express your answer as a common fraction.

    Solution: For the sum of the numbers to be even, either both are even, or both are odd. So, since there is a $\frac{1}{2}$ chance of getting an odd number and a $\frac{1}{2}$ chance for an even number on each die, there is a $\frac{1}{2} \times \frac{1}{2} = \frac{1}{4}$ chance of getting 2 even numbers, and a $\frac{1}{4}$ chance of getting 2 odd numbers. So, there is a $\frac{1}{4} + \frac{1}{4} = \boxed{\frac{1}{2}}$ chance of rolling an even sum.

20. In the month of April there are 30 days. How many possible ratios are there of the number of Mondays to the number of Thursdays?

    Solution: The most straightforward way to solve this problem is to bash out the calendar for each of the 7 possible arrangements. We see that if the month starts on Monday or Sunday, there will be 5 Mondays and 4 Thursdays, with a ratio of $\frac{5}{4}$. If April starts on Tuesday, Friday, or Saturday, there are 4 Mondays and 4 Thursdays, with a ratio of 1. If the first day is on a Wednesday or Thursday, 4 Mondays and 5 Thursdays produce the ratio $\frac{4}{5}$. Therefore, there are $\boxed{3}$ possible ratios.

21. $\triangle ABC$ is a right triangle with integer side lengths and hypotenuse BC. Point D is chosen such that A and D are on opposite sides of BC and $BD = CD = 8$. If $BC = 10$, find the perimeter of $ABDC$.

    Solution: We see that to find the perimeter of $ABDC$, we wish to find $AB + BD + DC + CA$. Recall that the only right triangle with integer sides and hypotenuse 10 is the 6-8-10 triangle. Therefore, $AB + BD + DC + CA = (AB + AC) + (BD + CD) = (6 + 8) + (8 + 8) = \boxed{30}$.

22. If Sally can wash a car in 1 hour, and Victor and Sally together can wash a car in 20 minutes, how long does it take Victor to wash a car by himself?

Solution: In 20 minutes, Sally washes $\frac{1}{3}$ of the car. So Victor must wash $\frac{2}{3}$ of the car in 20 minutes, which makes his total time $\boxed{30}$ minutes.

23. There are 498 students in your grade. It is time to take a class field trip to the Smithsonian Museums, and your teachers attempt to divide the class into equal groups such that each group has only one teacher. However, there are some students that will not be able to be placed into groups. If there are 24 teachers, at most how many students can be in a group?

Solution: We want the groups to be equal, but there will be some students remaining who have no teacher. Thus, the maximum number of students in a group is $\left\lfloor \dfrac{498}{24} \right\rfloor = \boxed{20}$.

24. A wood block has a density of 1.5 grams per cubic centimeter. Given that density is mass divided by volume, what is the mass, in grams, of 200 cubic centimeters of this material?

Solution: Since density is mass divided by volume, mass is density times volume. So our answer is $1.5 \times 200 = \boxed{300}$ grams.

25. Andy picked 6 apples from his apple tree. He cut each apple into 8 slices. When Andy wasn't looking, Alexis came and ate 15 apple slices. If Andy had been planning to give out the apple slices equally among 3 people, how many fewer slices does each person get?

   Solution: Since 15 apple slices were taken, $\frac{15}{3} = \boxed{5}$ were taken from each person.

26. A car traveling 95 feet per second is traveling how many miles per hour? There are 5280 feet in 1 mile. Express your answer to the nearest whole number.

   Solution: 95 feet per second is equivalent to $95 \cdot 3600 = 342000$ feet per hour which is equivalent to $\frac{342000}{5280} \approx \boxed{65}$ miles per hour.

27. Daisy and Jane are racing. If Jane gets a head start of 5 seconds, and Daisy catches up to Jane after running 60 feet in 10 seconds, then what was Jane's average speed in feet per second?

   Solution: Since Daisy catches up to Jane after running for 10 seconds, Jane has been running for $5 + 10 = 15$ seconds, because Jane had a 5-second head start. We see that Daisy catches up to Jane after running 60 feet, so Jane is 60 feet from the starting point. Thus, Jane runs at $60 \div 15 = \boxed{4}$ feet per second.

28. A list of 300 numbers starts with 1. After that, every number is three more than the number before it. What is the 200th number on the list?

   Solution: This is equivalent to finding the 200th number in the arithmetic progression with first term 1 and difference 3. Recall that this would be $1 + 3 \cdot (200 - 1) = \boxed{598}$.

29. If a is 50% of b, and b is 30% of c, what is the value of $\frac{a}{c}$? Express your answer as a common fraction.

Solution: a is 50% of 30% of c which is 15% $= \boxed{\dfrac{3}{20}}$.

30. Allen and Bob have pieces of candy. If Allen gave Bob 3 of his candies, the two boys would have the same amount of candy. Given that the boys have 40 pieces of candy in total, how much candy does Bob have?

Solution: If Allen gave Bob 3 of his candies, they would have 20 each. So Bob has $20 - 3 = \boxed{17}$ candies.

31. The equation used to convert C degrees Celsius to F degrees Fahrenheit is $F = \frac{9}{5}C + 32$. How many degrees Celsius is 113 degrees Fahrenheit?

Solution: We have $113 = \frac{9}{5}C + 32 \rightarrow 81 = \frac{9}{5}C \rightarrow C = \boxed{45}$.

32. Alice went into a fruit store and bought 21 pieces of fruit. The fruit store sells only apples and pears. Apples cost 12 cents each while pears cost 8 cents each. If her total cost was $2.12, how many apples did Alice buy?

Solution: If Alice bought a apples, she bought $21 - a$ pears. So we have $12a + 8(21 - a) = 212 \rightarrow 4a + 168 = 212 \rightarrow a = \boxed{11}$.

33. What is the value of $99 - 98 + 97 - 96 + 95 - 94 + \ldots + 3 - 2 + 1$?

Solution: $99 - 98 + 97 - 96 + 95 - 94 + \cdots + 3 - 2 + 1 = (99 - 98) + (97 - 96) + (95 - 94) + \cdots + (3 - 2) + 1 = 1 + 1 + 1 + \cdots + 1 + 1$. There are 49 grouped pairs such as $(99 - 98)$, so there are 50 1's, so $99 - 98 + 97 - 96 + 95 - 94 + \cdots + 3 - 2 + 1 = \boxed{50}$.

34. How many ways are there to arrange 5 different keys on a keychain? Two arrangements are considered the same if one can reach the other through rotation and/or reflections.

   Solution: Since rotations do not count, we fix a random keychain and arrange the 4 remaining ones around them. Since reflections do not count, we divide by 2. Hence, the answer is $4!/2 = \boxed{12}$.

35. The 600 students at a middle school are divided into three groups of equal size for lunch. Each group has lunch at a different time. A computer randomly assigns each student to one of three lunch groups. What is the probability that three friends, Kevin, Francis, and William, will be assigned to the same lunch group? Express your answer as a common fraction.

   Solution: Let the group Kevin is assigned to be called A. Francis then has a $\frac{1}{3}$ chance to be assigned to A, and William also has a $\frac{1}{3}$ chance. So, the chance that they are assigned to the same group is $\frac{1}{3} \cdot \frac{1}{3} = \boxed{\dfrac{1}{9}}$.

36. Compute: $\dfrac{13 + 26 + 39 + \ldots + 1300}{17 + 34 + 51 + \ldots + 1700}$.

   Solution: Notice that $\frac{13}{17} = \frac{26}{34} = \frac{39}{51} = \cdots = \frac{1300}{1700}$. Since $\frac{a}{b} = \frac{c}{d} = \frac{a+c}{b+d}$, all of this is just equal to $\boxed{\dfrac{13}{17}}$.

37. Eli spends 20% of his salary on food and spends 32% of his salary on housing. If he spends $585 on food and housing, what is Eli's salary (in dollars)?

   Solution: We have that Eli spends 52% of his salary on food and housing. So if Eli's salary is s, we have $\frac{52}{100}s = 585$. Now we simply solve for s, which comes out to be $\boxed{1125}$.

38. Billy Bob has 10 songs on his jPod Nano, 90 songs on his jPhone, and 30 songs on his jPod Touch. The 10 songs in his jPod Nano are also included in the 30 songs in his jPod Touch and the 90 songs on his jPhone. How many distinct songs does Billy Bob have in all of the devices?

Solution: The 10 songs on Billy Bob's jPod are repeated 3 times, so adding 10, 30, and 90 together means you overcount the number of songs by $2 \times 10 = 20$ songs. So, in total, Billy Bob has $10 + 30 + 90 - 20 = \boxed{110}$ songs in total.

39. There are 20 chapters in the math book *Math*, and the nth chapter corresponds with the index number of the nth odd number. For example, the first chapter's index number is 1, second chapter is 3, third chapter is 5, and so on. There are as many subsections in each chapter as the corresponding index number. How many subsections are there in the book?

Solution: The first chapter must have 1 subsection, the second has 3 subsections, the third has 5 subsections, and so on, so the twentieth has 39 subsections. So, there are a total of $1 + 3 + 5 + \cdots + 39 = \frac{(1+39) \times (20)}{2} = \boxed{400}$ subsections.

40. There are 900 students in school. 720 take Spanish. Three-fourths of the remaining students take French and the rest take Latin. How many students take Latin?

Solution: There are $900 - 720 = 180$ students who do not take Spanish, and Latin students are one-fourth of that, which is $\boxed{45}$.

41. Erik can finish a 20-question quiz in 10 minutes. Kathy can finish the same quiz in 15 minutes. If the two of them work together on the same 20-question quiz, in minutes, how long will it take them to finish?

Solution: Erik does 2 problems per minute, and Kathy does $\frac{4}{3}$ problems per minute. Together they do $\frac{10}{3}$ problems per minute, so it takes them $\frac{20}{\frac{10}{3}} = \boxed{6}$ minutes to do the test together.

42. If $x + y + z = 8$, where x, y, and z are nonnegative integers, compute the number of ordered triples (x, y, z).

Solution: This is Stars and Bars, for 3 spaces and 8 balls where each ball is equal to 1 in the 8. Thus, there are $\binom{8+3-1}{8} = \binom{10}{8} = \boxed{45}$.

43. What is the ratio of the number of 9 digit palindromes to the number of 8 digit palindromes? Recall that palindromes are numbers that read the same forward as backwards, and cannot have leading zeroes.

Solution: Notice that each 8-digit palindrome corresponds to exactly 10 9-digit palindromes: the one's found by inserting any digit at the center of the palindrome. Notice further that over all 8-digit palindromes, this accounts for every single 9-digit palindrome. Hence, the ratio is $\boxed{10 : 1}$.

44. A particular brand of lemonade is composed of 75% lemon juice and 25% water. I take 8 oz. of this lemonade and pour it into a beaker with 2 oz. of fresh water. What percentage of the fluid in the beaker is lemon juice?

Solution: 8 oz. of that lemonade contains 6 oz. of lemon juice and 2 oz. of water. The beaker will contain 6 oz. of lemonade and 4 oz. of water. So there will be 6 oz. of lemon juice out of 10 total ounces, which is $\boxed{60\%}$.

45. Sarah is delivering newspapers on a street of houses numbered 1 to 100 with even-numbered houses on the right and odd-numbered houses on the left. If she only delivers to house numbers that are a multiple of 3 and on the right, how many houses does she deliver newspapers to?

Solution: We see that Sarah only delivers to house numbers that are a multiple of 3 and even, since only evens are on the right. Thus, Sarah delivers to only the house numbers that are a multiple of 6. Therefore, Sarah delivers to $\left\lfloor \dfrac{100}{6} \right\rfloor = \boxed{16}$ houses.

46. Paul is not very good at talking to girls. Every time Paul tries to talk to a girl, he gets very nervous and starts counting in multiples of 3. If Paul counts in multiples of 3 (3, 6, 9, 12, 15,...) until he reaches 1,242, how many numbers has Paul said?

Solution: Since Paul counts in multiples of 3, to find the number of numbers that he has said when Paul reaches 1,242, we simply find $\dfrac{1242}{3} = \boxed{414}$.

47. Ryan ran 100 meters from the start line to the finish line in 15 seconds. He then ran back (from finish line to start line) at a pace of 5 meters per second. What was his average speed in meters per second of his 200 meter run? Express your answer as a decimal rounded to the nearest tenth.

Solution: When Ryan ran back to the starting line, he took $\frac{100}{5} = 20$ seconds. So in total he has taken 35 seconds to run 200 meters. That is an average speed of $\frac{200}{35} \approx \boxed{5.7}$ meters per second.

48. Alice, Allison and Ali are all siblings. Alice is twice as old as Ali, and the sum of the ages of Alice and Allison is 20. If Ali is 8, how old is Allison?

Solution: Alice must be 16, since she is twice as old as Ali, so Allison must be $\boxed{4}$.

49. Five mean monkeys are jumping on a bed. After every 10 minutes, the monkeys on the bed randomly and unanimously decide to kill one of their group. What is the probability that monkey Bob survives? Express your answer as a common fraction.

    Solution: After the first 10 minutes, the probability that Bob is not killed is $\frac{4}{5}$. After the second 10 minutes, the probability that Bob is not killed is $\frac{3}{4}$. After the third 10 minutes, the probability that Bob is not killed is $\frac{2}{3}$. After the fourth 10 minutes, the probability that Bob is not killed and is the last monkey alive is $\frac{1}{2}$. So, the probability that Bob survives is $\frac{4}{5} \times \frac{3}{4} \times \frac{2}{3} \times \frac{1}{2} = \boxed{\frac{1}{5}}$.

50. A point (x, y) is a *lattice* point if both x and y are integers. Given a *lattice* point (x, y), its *adjacent* lattice points are $(x + 1, y)$, $(x - 1, y)$, $(x, y + 1)$, and $(x, y - 1)$. A grasshopper starts at $(0, 0)$ on a coordinate plane. Each second, he jumps to an adjacent lattice point. How many possible points can the grasshopper move through or reach after 6 seconds?

    Solution: The sum of the coordinates must go either up or down by 1 each time the grasshopper moves. So, the sum of the final coordinates after 6 moves must be even, so the x and y coordinates must be either both even or both odd. The coordinates must also be between -6 and 6. If you mark all the possible lattice points, you get a 7 by 7 square of points tilted by $45°$ around the origin, which contains $\boxed{49}$ points.

51. Simplify the expression $\frac{2^2 \cdot \sqrt{4^3}}{12^2}$. Express your answer as a common fraction.

    Solution: $\sqrt{4^3} = \sqrt{64} = 8$. So we now have $\frac{4 \cdot 8}{144} = \frac{32}{144} = \boxed{\frac{2}{9}}$.

52. If an ant crawls around the outside of a square with side
length 1 inch, always keeping 1 inch away from the boundary
of the square, then how many inches will the ant travel in one
rotation around the square? Express your answer in terms of
π.

Solution: We see that when the ant is traveling parallel to
a side of the square from one corner to another, it travels the
length of a side of the square. When turning the corner, the
ant traces out a quarter circle with the corner as the center
and a radius of 1 inch. Therefore, the total perimeter of this
figure is $4 \cdot 1 + 2\pi \cdot 1 = \boxed{4 + 2\pi}$.

53. A turkey escapes from Holly's farm and
starts running away in a straight line at 4
meters per second. 10 seconds later, Holly
notices the turkey is missing and chases after
it at 8 meters per second. How far will Holly
run before she catches up to the turkey?

Solution: The turkey will have a 40 meter head-start. Holly
will have to take $\frac{40}{8-4} = 10$ seconds to catch up, within which
she will run $10 \cdot 8 = \boxed{80}$ meters.

54. Peter can solve *The Last Olympiad* problems in 3 hours, and
Annabelle can solve all of them in 4 hours. How many hours
will it take Peter and Annabelle to finish all the problems
in *The Last Olympiad*? Express your answer as a common
fraction.

Solution: Peter finishes $\frac{1}{3}$ of the problems every hour, and
Annabelle finishes $\frac{1}{4}$ of the problems every hour. Together,
they solve $\frac{1}{3} + \frac{1}{4} = \frac{7}{12}$ of the problems every hour. So it will
take them $\dfrac{1}{\frac{7}{12}} = \boxed{\dfrac{12}{7}}$ hours to finish the book.

55. Define the function $a\#b$ as $2a + 2b$. If $x + y + z = 3$, what is the value of $\left(\frac{1}{2}(x\#y)\right)\#z$?

Solution: First, we evaluate the expression within the parentheses. $\left(\frac{1}{2}(x\#y)\right) = \frac{1}{2}(2x + 2y) = x + y$. Now we have $(x+y)\#z$. This is equal to $(2x+2y)+2z = 2(x+y+z) = \boxed{6}$.

56. An astronaut goes to an alien planet where half of the citizens always tell the truth, and the others always lie. The astronaut meets three citizens, and asks the first one if the second one is a liar. It says, "No, it is not". Then the astronaut turns and ask the second one if the third one is a truth-teller. It says "No, it is not". Finally, the astronaut asks the third one if it is a liar. The third one says, "No, I am not". A truth-teller walks by and says, "1 lie has been told." What are each of the three citizens?

Solution: If 1 was a liar, then 2 would be a liar, and 3 would be a truth-teller. But then 2 lies would have been told. Therefore, 1 is a truth-teller, which makes 2 a truth-teller and 3 a liar.

57. Todd lives in a magical pond, where all the toads and frogs talk. One species always says the truth, while the other species always lies. Figure out the species of each talker.

Todd: My neighbor Tammy is a frog.

Tammy: My neighbors Todd and Ted are both toads.

Ted: I am a frog or a toad.

Terry: Todd is a toad.

Solution: If Todd is a frog, then Tammy must be of the lying species. If Tammy is a toad, then Todd would be lying, which would mean that both species lie. So then Tammy is a frog. But then Todd would be telling the truth and Tammy would be lying, even though they are the same species. So Todd is a toad. If toads tell the truth, then Tammy must be a frog, and frogs must lie. Thus Ted would have to be a frog, but Ted is telling the truth. Therefore, toads lie. The rest follows. Tammy is a toad, which makes Ted a frog, which makes Terry a frog because he tells the truth.

58. Maddy wants to buy school supplies. She visits a store that sells pencils in packs of seven, pens in packs of six, and erasers in packs of four. She wants to buy the same number of pencils, pens, and erasers, but can only purchase packs of them. What is the least possible number of pencils Maddy needs to buy?

Solution: This problem is asking for the LCM of seven, six, and four, which is $\boxed{84}$.

59. How many two-digit numbers are divisible by 11 and not even?

Solution: Out of the nine numbers $11, 22, \ldots, 99$, we cannot have four of them, namely $22, 44, 66, 88$. So our answer is $9 - 4 = \boxed{5}$.

60. John's mom is 30 years older than John. In 9 years, John's mom will be three times older than John. How old is John's mom now?

Solution: Let John's age be J, and his mom's age be M. We have $M = J + 30$, and $M + 9 = 3(J + 9) = 3J + 27$. So, $J + 30 = (M + 9) - 9 = 3J + 18$, and $2J = 12$, so John is 6 years old. John's mom is 30 years older than John, so her age is $30 + 6 = \boxed{36}$.

61. The distance between town A and town B is 150 miles. Andy is driving from A to B in a car that goes 60 miles per hour. However, there is a 30 mile stretch of road that is jammed with traffic. Traffic decreases the speed of the car by 80 percent. How long will it take Andy to get there, and what will be his average speed?

Solution: Andy will drive 120 miles at 60 miles per hour, and 50 miles at 20 percent of that speed. It will take him

$\frac{120}{60}$ hours = 2 hours to do the first one, and $\frac{30}{12}$ hours = 2 hours and 30 minutes. So it takes him a total of 4 hours and 30 minutes to get there, which makes his average speed $\frac{150}{4.5} \approx \boxed{33.3}$ mph.

62. Mister, Miss, and Junior Rabbit are eating out of a rabbit garden with 240 carrots in it. Mister Rabbit eats 24 carrots per hour. Miss Rabbit eats 36 carrots per hour. Junior Rabbit eats 20 carrots per hour. All rabbits eat at only half their speed after noon. If the three rabbits start eating out of the rabbit garden at 10 : 00 AM, by what time will they have finished all the carrots in the garden?

Solution: Combined, they all eat $24 + 36 + 20 = 80$ carrots per hour. So in the first two hours before noon, they eat 160 carrots. After noon, their rate goes down by half to 40 carrots per hour. Since they will have $240 - 160 = 80$ carrots left, they take two more hours after noon, which gives us the answer of $\boxed{2 : 00}$ PM.

63. Daniel has 20 coins, all of which are either pennies or nickels. If the total amount of Daniel's coins is \$0.32, how many nickels does Daniel have?

Solution: If Daniel has n nickels and p pennies, he has $5n + p$ cents. So we have $5n + p = 32$ and $n + p = 20$. Subtracting the second equation from the first gives us $4n = 12 \rightarrow n = \boxed{3}$.

64. In a xy-plane, let A be the point $(1, 1)$, B be the point $(3, 5)$, and C be the point $(4, 1)$. Point B is reflected across the x-axis to produce Point D. What is the area of $\triangle ACD$ in square units?

Solution: Point D is $(3, -5)$. Segment AC has length 3, and the altitude to D has length 6. So the area of $\triangle ACD$ is $\boxed{9}$.

65. City A is 11 miles from City B. City C is directly south of City B and east of City A. If the distance from City A to City C is 7 miles, what is the distance from City B to City C? Express your answer in simplest radical form.

Solution: Cities A, B, and C form a right triangle with A and B as the ends of the hypotenuse. A simple Pythagorean calculation yields $\sqrt{121 - 49} = \sqrt{72} = \boxed{6\sqrt{2}}$ as our answer.

66. Eugene has a jar of coins that consists of only nickels, dimes, and quarters. He has 24 nickels and dimes, 30 dimes and quarters, and 32 nickels and quarters. How many coins does he have in his jar?

Solution: Let n be the number of nickels, d be the number of dimes, and q be the number of quarters. We have $n + d = 24$, $d + q = 30$, and $n + q = 32$. To find the total number of coins, or $n + d + q$, we add the three equations together to get $2n + 2d + 2q = 86 \rightarrow n + d + q = \boxed{43}$.

67. Compute the number of ways to order the integers from 1 to 12 such that the even numbers must be in ascending order and the odd numbers must be in descending order.

Solution: Due to the condition that the even numbers must be ascending and the odd numbers must be descending, there is only one way to order each of the odds and the evens: 2, 4, 6, 8, 10, 12 and 11, 9, 7, 5, 3, 1. As such, we need only consider the positions of the odd and even numbers relative to each other, and we would know how to order the numbers using the two above sequences. This is equivalent to ordering 6 O's and 6 E's in a line, or $\binom{12}{6} = \boxed{924}$.

68. Three animals are arguing about a missing cookie. Frog says that Rabbit ate it, Rabbit says that Frog ate it, and Bird says that Frog is innocent. If exactly 2 animals are lying, who is sure to be innocent?

Solution: Clearly, we see that at least one of Frog and Rabbit is lying, or else both ate the cookie. If only one of them is lying, then Bird must be lying. Thus, Frog ate the cookie, and Rabbit is innocent. If both Frog and Rabbit are lying, then Bird must have eaten the cookie, and Rabbit is again innocent. These are the only two cases, so $\boxed{\text{Rabbit}}$ is sure to be innocent.

69. Farmer Jones raises chickens and donkeys. Chickens have 2 legs, and donkeys have 4 legs. If Farmer Jones owns 80 animals and the animals have a total of 232 legs, how many chickens does he own?

    Solution: If Farmer Jones has c chickens and d donkeys, we have $c + d = 80$ and $2c + 4d = 232$. If we multiply the first equation by 4 and subtract the second equation from it, we get $2c = 88 \rightarrow c = \boxed{44}$.

70. A fish tank is 50% full. Its length is twice its width and its height is twice its length. If the length is 8 inches, what is the volume of the water (in cubic inches)?

    Solution: The width is half the length, which is 4 inches, and the height is twice the length, which is 16 inches. So the volume of the tank is $4 \cdot 8 \cdot 16 = 512$ cubic inches. Since the tank is only half full, the volume of the water is $\boxed{256}$ cubic inches.

71. 3 inch long squares are cut out of the corners of a rectangle with dimensions of 11 inches by 14 inches. If the resulting net is folded to make a box without a lid, what is the volume of this box (in cubic inches)?

    Solution: The base of the box is 5 inches by 8 inches, since each side of the rectangle was reduced by 6. The height of the box is 3 inches, so our volume is $5 \cdot 8 \cdot 3 = \boxed{120}$ cubic inches.

72. If a regular hexagon and an equilateral triangle have the same perimeter, what is the ratio of the area of the hexagon to that of the triangle? Express your answer as a common fraction.

Solution: Let the side length of the hexagon be x. Then, the perimeter is $6x$, and the side of the equilateral triangle must be $2x$. The area of a regular hexagon is $\frac{3\sqrt{3}}{2}s^2$, where s is the side length, and the area of an equilateral triangle is $\frac{\sqrt{3}}{4}s^2$. So, the area of the hexagon in this problem is $\frac{3\sqrt{3}}{2}x^2$, and the area of the triangle is $\frac{\sqrt{3}}{4}(2x)^2 = \sqrt{3}x^2$. So, the ratio of the area of the hexagon to the area of the triangle is
$$\frac{\frac{3\sqrt{3}}{2}x^2}{\sqrt{3}x^2} = \boxed{\frac{3}{2}}.$$

73. A right circular cone has a radius of length r centimeters. It has a height of 6 centimeters and the length of its slant height is $r + 2$ centimeters. What is that value of r?

Solution: We have a right triangle with legs r and 6 and hypotenuse $r + 2$. Pythagorean Theorem gives us $r^2 + 6^2 = (r+2)^2 \rightarrow r^2 + 36 = r^2 + 4r + 4 \rightarrow 4r = 32 \rightarrow r = \boxed{8}$.

74. Mike's bicycle has a front wheel of diameter 2 feet and a rear wheel of radius 9 inches. When Mike rides his bike to the grocery store, his front wheel goes through 300 revolutions. How many revolutions does his rear wheel go through?

Solution: The front wheel has a circumference of 2π feet which equals 24π inches. The back wheel has a circumference of 18π inches. Since the wheels travel over the same distance, the proportion of revolutions they go through is the reciprocal of the proportion of the circumferences. Since the proportion of the circumferences is $\frac{24\pi}{18\pi} = \frac{4}{3}$, the proportion of revolutions is $\frac{3}{4}$. So for that proportion to hold, the back wheel must have gone through $\boxed{400}$ revolutions.

75. What are the rightmost two digits of 5^{555}?

Solution: For any power of 5 bigger than 25, the last two digits are always $\boxed{25}$.

76. Jane the macaw is flying through the rainforest. She spots 100 frogs on her 2 hour flight. If she never saw two frogs within 30 seconds or shorter, what is the longest time she could have flown without seeing a frog?

Solution: Jane could have seen 100 frogs in 50 minutes, based on the condition that she never saw 2 within 30 seconds. That leaves $\boxed{70}$ minutes for her to fly in peace.

77. Inside a circle are three smaller equal-sized circles of radius 6, where each is tangent to the outer circle and to the other two small circles. What is the radius of the large circle? Express your answer in the form $a + b\sqrt{c}$, where a, b, and c are integers, and c isn't divisible by the square of any prime.

Solution: Connect the centers of the smaller circles to form an equilateral triangle, as shown in the diagram below.

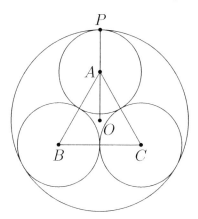

Since ABC is an equilateral triangle with side length 12, OA is two-thirds of the altitude, which is $6\sqrt{3}$. So, $OA = 4\sqrt{3}$, and $r = OP = OA + PA = \boxed{6 + 4\sqrt{3}}$.

78. If the radius of a cylinder is increased by 100%, how much should the height be decreased to maintain the same volume?

Solution: An increase of 100% means doubling the radius. Since the volume of a cylinder equals $\pi r^2 \cdot h$, doubling the radius would increase the volume by a factor of 4 if the height isn't changed. So, to keep the volume constant, the height must be decreased by a factor of 4, which means a $\boxed{75\%}$ decrease.

79. Two sets, A and B, are the same size. Their intersection consists of 100 elements, and their union has size 1000. Find the number of elements in set A.

Solution: Let the number of elements in set A be x. Then the number of elements in the union of A and B is equivalent to the number of elements in A plus the number of elements in B, minus the intersection. Thus, $x + x - 100 = 1000$, producing $x = \boxed{550}$.

80. Jim has 27 white 1×1 cubes. He decides to paint a single face of 10 cubes, 2 adjacent faces of 6 other cubes, and 2 opposite faces of 6 more cubes. When he assembles these cubes into a single 3×3 cube, what is the maximum fraction of its outside surface that will be painted?

Solution: For each of the 10 cubes that only has a single face painted, the maximum number of painted faces each can contribute to the outside surface is 1. For each of the 6 cubes with 2 adjacent faces painted, these can be placed at corners or edges of the large cube. Thus, each contributes 2 painted faces to the outside surface. For each of the 6 cubes with 2 opposite faces painted, each cube can contribute only 1 painted face since both faces cannot appear on the surface of the larger cube. Thus, there is a total of $10 \cdot 1 + 6 \cdot 2 + 6 \cdot 1 = 28$ painted faces. This gives a fraction of $\dfrac{28 \cdot 1^2}{6 \cdot 3^2} = \boxed{\dfrac{14}{27}}$.

81. A trapezoid has parallel bases having lengths of 3 units and x units. The height of the trapezoid is 6 units. If the area of the trapezoid is 30 square units, compute the value of x.

Solution: The area of a trapezoid is the product of the average of the bases and the height. So we get that the average of the bases is 5, which means $\frac{3+x}{2} = 5 \to x = \boxed{7}$.

82. What is the area of a triangle with vertices at $(3, 4)$, $(7, 9)$, and $(18, 8)$?

By the Shoelace Theorem, the area is equal to the absolute value of

$$\frac{\begin{vmatrix} 3 & 4 & 1 \\ 7 & 9 & 1 \\ 18 & 8 & 1 \end{vmatrix}}{2} = \boxed{\frac{59}{2}}$$

83. A robot is standing on a conveyor belt which moves to the left at a constant rate of 5 feet per second. The robot is programmed to walk to the right at 10 feet per second. If the conveyor belt is 110 feet long and the robot starts on the farthest left side of the conveyor belt, how many seconds will it take for the robot reach to the other side of the conveyor belt?

Solution: If the robot walks to the right at 10 feet per second on a conveyor belt that moves him to the left at 5 feet per second, his total movement will be 5 feet per second to the right. It will take the robot $\frac{110}{5} = \boxed{22}$ seconds to walk to the other side.

243

Serious Challenges

1. Let $P(x)$ be a polynomial of degree 4. If $P(0) = \frac{1}{2}$, $P(1) = \frac{2}{3}$, $P(2) = \frac{3}{4}$, $P(3) = \frac{4}{5}$, and $P(4) = \frac{5}{6}$, compute $P(5)$.

   Solution: Note that $P(x) = \frac{x+1}{x+2}$, or $(x+2)P(x) - (x+1) = 0$ for $x = 0, 1, 2, 3, 4$. Let $Q(x) = (x+2)P(x) - (x+1)$. The degree of P is 4 so the degree of Q is 5. Since a polynomial of degree 5 has 5 roots and $Q(0) = Q(1) = Q(2) = Q(3) = Q(4) = 0$, we know that $Q(x) = a(x)(x-1)(x-2)(x-3)(x-4)$ for some number a. Notice that $Q(-2) = 0 \times P(-2) - (-1) = 1$. Also, $Q(-2) = a(-2)(-3)(-4)(-5)(-6) = -720a$. Thus, $a = -1/720$ and $Q(x) = -1/720(x)(x-1)(x-2)(x-3)(x-4)$. Finally, $Q(5) = \frac{-5 \times 4 \times 3 \times 2 \times 1}{720} = -\frac{1}{6}$ and $Q(5) = 7P(5) - 6$ so $7P(5) - 6 = -\frac{1}{6}$ and $P(5) = \frac{5}{6}$.

2. How many ways are there to distribute 12 pieces of candy to 5 children? Not every child has to get candy, but all the candy must be distributed.

   Solution: The pieces of candy can be represented like this: OOOOOOOOOOOO. Then, four lines can be drawn in between groups of candy: O|OOO|OO|OOOO|OO. These four lines separate the 12 pieces of candy into 5 groups, which represent the pieces of candy each child gets. A child not getting any candy is represented by putting two lines right next to each other. For example, OOOOOOOOOOOO|||| delineates one child getting all the candy. So, every different ordering of 12 O's and 4 lines will give a new way to distribute the candy. There are $\binom{16}{4} = \frac{16!}{12!4!} = 1820$ ways to order the 12 O's and 4 lines, producing $\boxed{1820}$ ways to distribute the candy.

3. ABC is an acute triangle. Let O and H denote the circumcenter and orthocenter of the triangle respectively. If $\angle BAH = 35$ degrees, compute the value of $\angle CAO$ (in degrees) (Note: The circumcenter of a triangle is the center of the circle that circumscibes the triangle. The orthocenter of a triangle is the point of intersection of the triangle's three altitudes.)

Solution: Since the circumcenter of the triangle is the center of the circle that circumscribes it, $\angle AOC = 2\angle ABC$. Since $\triangle AOC$ is isosceles, $\angle CAO = \angle ACO$ and $\angle CAO + \angle ACO = 180 - \angle AOC$, so $\angle CAO = \frac{1}{2}(180 - \angle AOC) = \frac{1}{2}(180 - 2\angle ABC)$. Since AH is the altitude from A to BC, if AH intersects BC at D, then $\triangle ADB$ is a right triangle, so $\angle ABC = 90 - \angle BAH = 55$ degrees. So, $\angle CAO = \frac{1}{2}(180 - 2\angle ABC) = \frac{1}{2}(180 - 110) = \boxed{35}$ degrees.

4. Compute the last two digits of 15^{99}.

Solution: We first find the tens digit. By the binomial theorem, $(15)^{99} = (10+5)^{99} = 10^{99} + \binom{99}{1}10^{98} \times 5 + \binom{99}{2}10^{97} \times 5^2 + \ldots + \binom{99}{97}10^2 \times 5^{97} + \binom{99}{98}10 \times 5^{98} + 5^{99}$. We only look at the last 2 terms as the rest are all divisible by 100. $\binom{99}{98}10 \times 5^{98} + 5^{99} = 99 \times 10 \times 5^{98} + 5^{99}$. For the first term, as 99×5^{98} ends in a 5, this contributes 5 to the ten's digit. For the second term, recall that 5 raised to any integer greater than 1 ends in 25, which has a 2 in the tens digit. Thus, $5 + 2 = 7$ is our tens digit. Finally, this power obviously ends in 5, so our answer is $\boxed{75}$.

5. Compute the last two digits of $9^{3^{7^{6^5}}}$.

Solution: We wish to find the last two digits of the number, so we want $9^{3^{7^{6^5}}}$ (mod 100). We apply Euler's Theorem. This indicates that $9^{3^{7^{6^5}}} \equiv 9^x$ (mod 100), where $x = 3^{7^{6^5}}$ (mod $\varphi(100)$). $\varphi(100)$ is the totient function of 100, so $\varphi(100) = (5^2 - 5^1) \cdot (2^2 - 2^1) = 40$. We then wish to calculate $3^{7^{6^5}}$ (mod 40). Using another application of Euler's Theorem, we find that $\varphi(40) = (2^3 - 2^2) \cdot (5^1 - 5^0) = 16$. Since $7^2 \equiv 1$ (mod 16), we find that $7^{6^5} \equiv 1$ (mod 16) as 7^{6^5} is a power of 7^2. Then $3^{7^{6^5}} \equiv 3^1 \equiv 3$ (mod 40), so $9^{3^{7^{6^5}}} \equiv 9^3 \equiv 729$ (mod 100). Thus the last two digits are $\boxed{29}$.

6. Given that a and b are integers such that $0 < a < b$ and $2ab + 3a + 7b = 100$, compute a.

Solution: This is a little different from a standard Simon's Favorite Factoring Trick problem. Multiplying both sides by 2, we have $4ab + 6a + 14b = 200$. Now, $(2a + 7)(2b + 3) = 221$. After some experimentation, we see that $221 = 13 \times 17$. Thus, $2b + 3 = 17$ and $2a + 7 = 13$ so $a = 3$ while $b = \boxed{7}$.

7. If $a^2 + 2a + b^2 + 4b + c^2 + 6c = -14$, compute $a^2 + b^2 + c^2$.

Solution: Complete the square for each variable to get $a^2 + 2a + 1 + b^2 + 4b + 4 + c^2 + 6c + 9 = 0$ and $(a + 1)^2 + (b + 2)^2 + (c + 3)^2 = 0$. Since the square of a number is always greater than or equal to 0, all these squares must equal 0, so $a = -1, b = -2$, and $c = -3$. Thus, the sum of the squares is $\boxed{14}$.

8. Yvone and Zach are playing a game. They each roll 2 dice. Whoever has the higher sum wins. If the sums are the same, Yvone wins. What is the probability that Yvone will win?

Solution: First, calculate the probability that their sums are tied. The probabilities of rolling the sums from 2 to 12 are 1, 2, 3, 4, 5, 6, 5, 4, 3, 2, and 1 thirty-sixths, respectively. Thus, the probability that Yvone and Zach roll the same sums is

$$\left(\frac{1}{36}\right)^2 + \left(\frac{2}{36}\right)^2 + \cdots + \left(\frac{5}{36}\right)^2 + \left(\frac{6}{36}\right)^2$$

$$+ \left(\frac{5}{36}\right)^2 + \left(\frac{4}{36}\right)^2 + \cdots + \left(\frac{1}{36}\right)^2$$

$$= \frac{1 + 4 + 9 + 16 + 25 + 36 + 25 + 16 + 9 + 4 + 1}{1296}$$

$$= \frac{146}{1296} = \frac{73}{648}$$

So, the probability that one of them will have a higher sum is $1 - \frac{73}{648} = \frac{575}{648}$. If one of them rolls a higher sum, half of the time Zach will win, and the other half of the time Yvone will win. Therefore, the probability Yvone will win is

$$\frac{73}{648} + \frac{1}{2} \cdot \frac{575}{648} = \frac{146}{1296} + \frac{575}{1296} = \boxed{\frac{721}{1296}}$$

9. A point P is chosen in the interior of square $ABCD$ such that $AP = 5, BP = 1$, and $DP = 7$. Compute the area of triangle ACP.

   Solution: It is not hard to show that $AP^2 + CP^2 = BP^2 + DP^2$. Hence, $CP = 5$. By the SSS congruence statement, triangles CDP and ADP are congruent. Hence, $\angle CDP = \angle ADP$ and thus BD is a diagonal of length 8. This implies that AC also has length 8. Let the intersection of the diagonals be O. Triangle AOP has a right angle at $\angle AOP$ and $AO = 4$ so $PO = 3$ and finally, the area of triangle ACP is $\boxed{12}$.

10. If $x^3 + 8x^2 + 6x + 3 = 0$ has roots a, b, c, compute $a^2 + b^2 + c^2$.

    Solution: By Vieta's Formulas, $a + b + c = -8$ and $ab + bc + ca = 6$. Hence, $a^2 + b^2 + c^2 + 2(ab + bc + ca) = 64$ from squaring the first equation and $a^2 + b^2 + c^2 = \boxed{52}$ from substitution.

11. Given $x+y+z = 1$, $x^2+y^2+z^2 = 89$, and $x^2yz+xy^2z+xyz^2 = -84$, compute x, y, and z, if $x \le y \le z$.

    Solution: We are given that $x+y+z = 1$. Squaring, we find that $(x + y + z)^2 = x^2 + y^2 + z^2 + 2(xy + xz + yz) = 1$. Thus, $xy+xz+yz = -44$. Furthermore, we find that $x^2yz+xy^2z+xyz^2 = xyz(x + y + z) = -84$, so $xyz = -84$. We construct a polynomial with these equations: $r^3 - r^2 - 44r + 84 = 0$. We find that the three zeros are 2, -7, and 6. Since $x \le y \le z$, we find that $\boxed{x = -7, \ y = 2, \ z = 6}$.

12. Let N be an integer with 33 digits, all of which are 1's except for the 17th digit. Find the value of the 17th digit if N is divisible by 13.

Solution: Recall that the integer $111,111$ is divisible by 13 as it is divisible by $1001 = 7 \cdot 11 \cdot 13$. Thus, a 33-digit integer with all digits as 1's is equivalent to $111 \equiv 7 \pmod{13}$. We then wish to add $x \cdot 10^{16}$ to create the 17th digit. Recall that $10^{12} \equiv 1 \pmod{13}$ by Fermat's Little Theorem, so $10^{16} \equiv 10^{12} \cdot 10^4 \equiv 10^4 \equiv 3 \pmod{13}$. Thus, we want to find an x so that $N \equiv 7 + x \cdot 3 \equiv 0 \pmod{13}$. $x = 2$ is the only solution such that the 17th digit is less than 10, so the answer is $\boxed{3}$.

13. What is the minimum value of $ab + bc + ca$ if $abc = 1$?

Solution: Using simple AM-GM, we get that $\frac{ab+bc+ca}{3} \geq \sqrt{(abc)^2} = 1$. So the minimum value of $ab + bc + ca$ is $\boxed{3}$.

14. There are two barrels that look identical from the outside. Barrel A contains 6 red balls and 6 green balls, while barrel B contains 10 red balls and 2 green balls. All of the balls are identical, besides their color. Mike randomly chooses one of the barrels, and randomly draws a ball. If the ball is red, what is the probability that he picked barrel A?

Solution: The probability Mike picked barrel A given that he drew a red ball is equal to the probability Mike picked barrel A and drew a red ball divided by the probability Mike drew a red ball.

The probability Mike picked barrel A and drew a red ball is

$$\frac{1}{2} \cdot \frac{6}{12} = \frac{1}{4}$$

The probability Mike drew a red ball is

$$\frac{1}{2} \cdot \frac{6}{12} + \frac{1}{2} \cdot \frac{10}{12} = \frac{1}{4} + \frac{5}{12} = \frac{2}{3}$$

As a result, the final probability is $\frac{\frac{1}{4}}{\frac{2}{3}} = \boxed{\frac{3}{8}}$.

15. A circle is inscribed in isosceles triangle ABC (with $AB = AC$) and is tangent to sides BC, CA, and AB at P, Q, and R, respectively. Triangle PQR is drawn. Point X is drawn on arc QR which does not contain P. If $\overline{AQ} = 3$ and $\overline{AB} = 9$, what is the value of $\dfrac{\overline{XP}}{\overline{XQ} + \overline{XR}}$?

Solution:

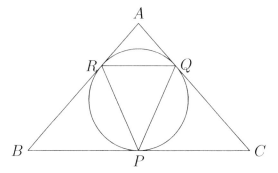

We see that RB $= 6$, and therefore BP $= 6$. So we get that BC $= 12$. By similar triangles, we see that QR $= 4$. We get that the altitude from A is $3\sqrt{5}$ so the distance between QR and BC is $2\sqrt{5}$. Dropping the altitude from Q to BC, we see by similar triangles that the distance from the foot of that altitude to C is 4. So the distance from the foot to P is 2. We then compute PQ $=$ PR $= 2\sqrt{6}$. Using Ptolemy's, we get that $\frac{XP}{XQ+XR} = \frac{PQ}{QR} = \boxed{\dfrac{\sqrt{6}}{2}}$.

16. Given a cubic with nonnegative roots p, q, and r, if $(p + q + r)^2 = 49$, $p^2 + q^2 + r^2 = 25$, and $pqr = 0$, what is $(x - p)(x - q)(x - r)$?

Solution: We get $2(pq + qr + pr) = 24$ by subtracting the second given from the first. Also, note that $p + q + r = \pm 7 \rightarrow p + q + r = 7$ because the roots are nonnegative. Then, using Vieta's, construct the polynomial as $x^3 - (p+q+r)x^2 + (pq + pr + qr)x - pqr = \boxed{x^3 - 7x^2 + 12x}$.

17. If the roots of the polynomial $x^4 - 5x^3 - 7x^2 + 29x + 30$ are a, b, c, and d, what is the value of $(a+1)(b+1)(c+1)(d+1)$?

Solution 1: The expression whose value we wish to compute is equivalent to $abcd + abc + bcd + cda + dab + ab + bc + cd + ac + ad + bd + a + b + c + d + 1$. By Vieta's, this is equal to $30 - 29 + (-7) - (-5) + 1 = \boxed{0}$.

Solution 2: Let $P(x) = x^4 - 5x^3 - 7x^2 + 29x + 30$. Note that we can write $P(x) = (x-a)(x-b)(x-c)(x-d)$, so

$$P(-1) = (-1-a)(-1-b)(-1-c)(-1-d)$$

$$P(-1) = (1+a)(1+b)(1+c)(1+d)$$

As a result, the answer is

$$P(-1) = (-1)^4 - 5(-1)^3 - 7(-1)^2 + 29(-1) + 30 = \boxed{0}$$

Orange County Math Circle: Where Math & Service Meet

(*Imagine* Magazine Vol. 18 No. 4)

Johnny (front, center) formed the Orange County Math Circle to involve young mathematicians in problem solving and community service. Here, OCMC volunteers gather at the 2010 Thanksgiving Math Tournament.

by Jonathan Li

Where Math & Service Meet

"Welcome to the 2010 Orange County All-Girls Math Tournament!" my voice boomed from the podium. My welcome was greeted with waves of high-pitched screams from 180 girls who flooded the gymnasium at St. Anne School. For a moment, I thought I had arrived at a Justin Bieber concert. But instead of sharing riveting pop music, I was here to spread the word of Fermat, Euler, and Gauss—and the young mathematicians in the audience couldn't have been more excited.

My path to that podium started many years earlier. After I participated in the Johns Hopkins Talent Search in fourth grade, my school allowed me to accelerate by taking high school math classes in fifth grade. Worried that I might run out of math classes, my mother searched for enrichment opportunities and found the San Diego Math Circle, a free problem-solving class for motivated, advanced math students held on Saturday mornings at UC San Diego.

I was in sixth grade when I attended for the first time, and I was captivated immediately. Mr. Richard Rusczyk, the founder of the San Diego Math Circle, taught us that true math is not memorizing formulas but implementing creative and original ideas. He taught us to approach problems with a combination of creativity, intuition, and discipline. As much as I loved his lectures, I also enjoyed socializing with 30 other like-minded students. Despite the long commute, attending the circle and sharing my solutions with my peers became one of my favorite activities. Later in the year, as my math skills improved, I also joined the Long Beach Math Circle, founded by Professor Kent Merryfield of Cal State University at Long Beach. Prof. Merryfield is also the head coach of the Southern

California American Regions Math League (ARML) Team, which trains top math students for the national ARML contest.

For the next three years, my weekends often involved driving 120 miles round trip to the San Diego Math Circle, or 110 miles round trip to the Long Beach Math Circle—and sometimes both in a single day.

Basic Training

By the time I entered high school in 2007, I was thinking seriously about bringing the math circle experience to my hometown in Orange County. I envisioned creating a student-run community-service organization where math enthusiasts could both explore advanced problem-solving topics and bring the math circle experience to underserved students in Southern California. By integrating learning and teaching, Orange County Math Circle (OCMC) members would work together to serve our community, strengthen math culture, and make math resources accessible to all.

my mentors and teachers to join the OCMC Advisory Board.

As my training progressed, I began offering monthly lectures on topics such as number theory, probability, and geometry to about 20 students from seven different schools. My first class, held at Concordia University in Irvine in October 2008, was designed to prepare students for the upcoming AMC 8 exam. As time went on, my program began to attract a group of loyal students who then invited their friends to the Math Circle, and I spread the word through my networks in math communities. By the end of the school year, my classes were regularly filled with about 40 students.

Expanding Opportunities

After running OCMC myself for two years, I focused on building it into a community-service organization that would attract more student volunteers. I asked my ARML team members to help and posted calls for volunteers on the team forum at ArtofProblemSolving.com. At the end of September 2009, with

The OCMC is unique in that it is run by students. Here, Johnny teaches a math circle class.

With no facilities, no volunteers, no students, no money, and no experience in running a math circle, the idea of starting one seemed daunting. But what I did have was my passion for math, enthusiasm to serve, and determination to work hard.

While researching how to start a community service organization, I learned that the Davidson Institute was launching its Davidson Young Scholars Ambassador Program. I had been named a Davidson Young Scholar in sixth grade, so I was eligible to apply. I submitted a proposal outlining my vision for OCMC and was accepted into the program. Through online seminars, discussion forums, and individualized mentoring, I spent 14 months learning skills such as goal setting and proposal writing, leadership, advocacy, fundraising, and public relations. Step by step, I applied what I learned to get OCMC started. I created a mission statement, wrote a business plan, and invited some of

11 volunteers from seven local high schools, we held our first organization meeting, where we decided to organize a math tournament: the Thanksgiving Math Tournament for students in grades three to six.

We decided to use MATHCOUNTS as our model, but added a few fun activities to make math a cool activity for younger students to enjoy. For example, we planned a Math Relay where teams of students would solve math problems in an outdoor relay format: a student from each team would run a 50-meter dash to a problem station to solve a problem and then return to tag the next student in line to continue. The first team to solve all 10 problems correctly would win the relay. I organized subcommittees for problem writing, grading, registration, promotion, and fundraising so volunteers could select activities that interested them. As more people learned about the tournament

After the 2010 All Girls Math Tournament, the OCMC saw a 150 percent increase in the number of girls participating in the math circle.

we attracted more volunteers. In the end, about 20 volunteers came out for the tournament, where more than 100 students spent a fun afternoon sharing the joy of mathematics.

Most of our volunteers had participated in MATHCOUNTS in middle school. MATHCOUNTS is a program for students in grades six to eight, but because most middle schools in Orange County have only grades seven and eight, most sixth graders don't get to participate. To offer younger students an opportunity to experience MATHCOUNTS while giving older students a chance to warm up for the real event, we created a MATHCOUNTS scrimmage.

The 2010 New Year's Invitational and MATHCOUNTS Scrimmage attracted about 100 students, and the Countdown Round was won by a fifth grader!

At that event, during a conversation with a group of parents, one mother told me, "My daughter is very good at math, but not too many girls at her school are interested." Over the years, I had noticed the small number of girls at all levels of math competitions. So for our next event, I proposed the 2010 All-Girls Math Tournament for girls in grades three through eight. Female OCMC volunteers overwhelmingly supported the idea, and many of them stepped up to take leadership for organizing the tournament. Highlights included a keynote speech by Dr. Natalia Komarova, a math professor from UC Irvine, and Lollipop Riddle Contests, where students worked in teams to solve math puzzles related to the number

Want to Join a Math Circle?

The National Association of Math Circles lists existing math circles in 31 states. Visit their website to see if there is a program near you: www.mathcircles.org/Wiki_ExistingMathCirclePrograms. Note that these programs have a variety of affiliations: many are affiliated with universities, some are based at schools, others are run by parents, and OCMC, of course, is run by students.

If no math circle exists in your area, consider doing what Johnny did: start your own! Circle in a Box, available in both print and PDF forms, offers suggestions for starting and sustaining a math circle, and even includes materials for math circle presentations, sample grant proposals, and templates for other administrative tasks.

Download the PDF
http://minerva.msri.org/files/circleinabox.pdf

Order the Book
www.ams.org/bookstore-getitem/item=mcl-2

Johnny works with students in the Santa Ana Math Club, which was created with the help of OCMC.

Math Circles: A Brief History

Mathematical enrichment activities in the United States have been around for at least 30 years, in the form of residential summer programs, math contests, and local school-based programs. The concept of a math circle, on the other hand, with its emphasis on convening professional mathematicians and secondary school students on a regular basis to solve problems, has appeared only within the past 12 years.

This form of mathematical outreach made its way to the U.S. most directly from Russia and Bulgaria, where it has been a fixture of their mathematical culture for decades. (The first ones appeared in Russia during the 1930s; they have existed in Bulgaria for a century.) The tradition arrived with emigres who had received their inspiration from math circles as teenagers. Many of them successfully climbed the academic ladder to secure positions within universities, and a few pioneers among

them decided to initiate math circles within their communities to preserve the tradition which had been so pivotal in their own formation as mathematicians. The Mathematical Sciences Research Institute (MSRI) in Berkeley, California, became involved at an early stage by supporting the Berkeley Math Circle. Not long after, Steve Olson highlighted this math circle in his book *Countdown*, since a couple of members of the 2001 U.S. International Mathematical Olympiad (IMO) team attributed their success in part to the problem-solving sessions offered at Berkeley. In this and other ways, math circles began to attract national attention as a means for encouraging students to enjoy, explore, and excel in mathematics.

—from *Circle in a Box* by Sam Vandervelde
(Mathematical Sciences Research Institute, 2007).

---

of lollipops in a bag. This tournament led to a 150 percent increase in girls' participation in our Math Circle program.

Our most recent undertaking has been to start the Santa Ana Math Club. While volunteering at the Santa Ana Math Field Day in June 2010, four OCMC volunteers and I had the opportunity to help over 400 low-income students participate in mathematical problem solving. After gaining support from the Santa Ana Unified School District, I wrote a proposal to the Mathematical Sciences Research Institute (MSRI) and was awarded a $2,000 grant to launch the project. Today, the Club has enrolled 84 students in grades four through seven, and each month, several OCMC volunteers spend a Saturday morning teaching classes to the members of the club.

Looking Back and Forward

Four years later, OCMC has evolved into a unique community for aspiring mathematicians. Our work pioneered a new model for math circles: in addition to providing opportunities for motivated pre-college students to explore advanced mathematics, we give them a way to become ambassadors who bring the beauty of mathematics to a broader audience.

Thanks to our more than 50 devoted volunteers from 30 schools, OCMC's lectures, tournaments, and special events have served more than 1,000 students from 75 schools. Our math tournaments have become annual events many students look forward to. When California's budget cuts forced schools to eliminate enrichment programs, OCMC became a

destination for many students whose schools could no longer offer math clubs. We assembled OCMC math teams through which such students can participate in math competitions. Last year, our team placed fourth in the inaugural Caltech Harvey Mudd Math Competition. Recently, one of our teams finished first nationally in the Team Round and third overall at the annual Harvard-MIT Online Math Tournament.

While I have been honored to win many national math and science awards, creating OCMC is what I've found most rewarding and what I'm most proud of. Through the process, I learned the importance of compassion and leadership in pursuit of math that serves humanity. As much as I want to continue to lead OCMC through its expansion and growth, I know that my role will change when I enter college in the fall. With that in mind, I have been spending more time preparing younger volunteers to step up to leadership roles in OCMC. And then I will join OCMC's Advisory Board, where I hope to mentor aspiring young mathematicians for years to come.   i

Jonathan Li is a senior at St. Margaret's Episcopal School in San Juan Capistrano, CA. He is a Davidson Fellow, an Intel STS Finalist, a United States Physics Team member, a Math Olympiad Summer Program (MOSP) participant, a three-time USAMO qualifier and the captain of the Southern California ARML Team. As a cellist, Jonathan has played in All-State and All-Southern Honor Orchestras. At school, he plays varsity soccer, serves on the Honor Committee, and heads up Mu Alpha Theta and JETS Teams. He will attend Harvard in the fall.

For more information:

Davidson Young Scholars Program
www.davidsongifted.org/youngscholars

Mathematical Sciences Research Institute
www.msri.org

National Association of Math Circles
www.mathcircles.org

Orange County Math Circle
www.ocmathcircle.org